MW00837886

Deep Learning Applications of Short-Range Radars

For a complete listing of the *Radar Series*, turn to the back of this book.

Deep Learning Applications of Short-Range Radars

Avik Santra

Souvik Hazra

ARTECH HOUSE

BOSTON | LONDON
artechhouse.com

Library of Congress Cataloging-in-Publication Data
A catalog record for this book is available from the U.S. Library of Congress.

British Library Cataloguing in Publication Data
A catalogue record for this book is available from the British Library.

Cover design by John Gomes

ISBN 13: 978-1-63081-746-6

10 9 8 7 6 5 4 3 2 1

Contents

Preface

Radar has evolved from a complex, high-end military technology into a relatively simple, low-end solution penetrating industrial, automotive, and consumer market segments. This rapid evolution has been driven by two main factors: advancements in silicon and packaging technology that has led to miniaturization, and growth of computing power that has enabled the use of deep learning algorithms to tap the full potential of radar signals. The use and applications of radar technology have grown multifold in recent years. From military and defense applications, radars have been widely used in automotive applications to increase safety and facilitate driving in medium- to premium-priced cars, and is also an important sensor for autonomous cars. For adoption of short-range radars for several industrial, consumer, and in-cabin automotive applications require reliable system performance at small-form factor, low power, and low cost.

The advent of deep learning has transformed many fields and resulted in state-of-the-art solutions in areas such as computer vision, natural language processing, and speech processing. However, the application of deep learning algorithms to radars is still by and large at its nascent stage. A radar system consists of two parts: first, the radar hardware, including the RF transceiver, waveform generator, receiver unit, antenna and system packaging. State-of-the-art SiGe and CMOS are candidate technologies for mm-wave short-range radars and offer flexibility for integration and smaller form-factor. The second part is the sensing aspect, which relies on signal processing or deep learning algorithms that parses the radar return echo into meaningful target information that facilitates a desired application. The goal of this book is to demonstrate and highlight how deep learning is enabling several advanced industrial, consumer, and in-cabin applications of short-range radars, which weren't otherwise possible. The book illustrates various advanced applications, their respective challenges, and how they are being addressed using different deep learning architectures and algorithms. While several applications presented in the book are at a high product readiness stage, some of them form the basis for further research and development for product readiness.

The book is laid out as follows: Chapter 1 introduces radar signal processing, the various types of short-range radars, the basics of waveform designs, target models for high-resolution radars, and 3D radar data-cube processing involving detection, parameter estimation, and tracking algorithms. The chapter concludes with various advanced industrial, consumer and automotive applications of short-range radars.

Chapter 2 provides an introduction to deep learning, outlining the history of neural networks and the optimization algorithms to train them. The chapter introduces the modern deep convolutional neural network (DCNN), popular DCNN architectures for computer vision is a field of study dealing with processing of images. It presents other deep learning architectures such as long-short term memory (LSTM), autoencoders, variational autoencoders (VAE), and generative adversarial networks (GANs).

Chapter 3 introduces the application of gesture recognition. We divide them into macrogesture sensing and recognition, which involve major arm movements and are performed from a distance to the sensor, as well as microgesture sensing and recognition, which involves minute subtle finger movements and are performed close to the sensor. While the macrogesture finds application in projectors or TVs, the microgesture finds applications in smartphones, smartwatches, and automotive dashboards.

Chapter 4 introduces the application of human activity recognition. The chapter presents various DCNN architectures for enabling activity recognition and comparing their performance. The chapter further introduces the concept of continuous activity recognition using a combination of tracking and a deep learning classifier such as LSTM. The chapter end by presents the topic of elderly fall motion recognition, the various challenges, and how it can be addressed through deep learning.

Chapter 5 presents the application of air-writing using a network of short-range radars. Air-writing or air-drawing refers to linguistic characters drawn on an imaginary board and the system recognizing these characters. Air-writing finds application in human-machine interface for augmented-reality virtual reality (AR-VR) and alternate interface mechanisms for desktops.

Chapter 6 presents the application of material classification. The objective of short-range radar is to classify among everyday objects and materials, and finds applications in vacuum cleaners and robots. The chapter concludes with the challenges to be overcome for integration into deployable system solutions.

Chapter 7 presents the application of remote vital sensing using short-range radars. The chapter presents various challenges with a pure deep learning solution and introduces a tracking-based solution as a stable solution, and also introduces a hybrid tracking with deep learning solution as a more reliable and scalable solution.

Chapter 8 introduces the application of short-range radars in conjunction with deep learning for human presence, counting, and localization applications. The chapter presents how deep learning has and is replacing the classical signal processing pipeline to improve system performance as well as addressing challenges that weren't feasible with a classical signal processing pipeline.

Chapter 9 introduces the application of in-cabin sensing. Apart from automatic cruise control (ACC), blind spot detection (BSD), and parking assistance, radars are enabling a plethora of in-cabin sensing applications such as smart trunk opening (STO) applications, child-left-behind applications, and smart airbag applications. We conclude the chapter with a federated learning framework, wherein a mechanism to automatically update a deployed deep learning model in production is outlined.

The book is intended for graduate students, academic researchers, and industry practioners working with mm-wave radars and radar algorithms. The book is written for beginners to advanced researchers and assumes sufficient knowledge of linear algebra and engineering mathematics. Each chapter has a question section to assess the understanding of the reader. The book covers conventional deep learning architectures, as well as the adaptations required for product-ready solution for an application. It covers advanced concepts like meta-learning, multimodal cross-learning, attention mechanism, federated learning, deformable convolution, and PointNets. While each chapter is independent of the others, it is suggested that an early researcher read the first two introductory chapters before reading any application-specific chapter. This book also provides an introduction to classical signal processing algorithms in Chapter 1 and basics of deep learning algorithms in Chapter 2.

The authors would like to thank their respective families: Avik Santra would like to thank his wife Sudarshana for her constant love, encouragement, and motivation that helped him shape the book, and his son Abhinava who is a perpetual source of energy and happiness. The authors would like to acknowledge and thank their manager, Thomas Finke, for his tremendous support and encouragement during the period of writing this book. The authors would also like to thank their colleagues and masters students, who worked together on various projects. In particular, the authors would like to thank Prachi Vaishnav, Raghavendran Vagarappan Ulaganathan, Christoph Will, Yogesh Shankar, Jonas Weiss, Michael Stephan, Muhammed Arsalan, and Rodrigo Hernangomez. The authors would also like to thank anonymous reviewers for their encouragement, reviews, and suggestions to improve the book.

1

Introduction to Radar Signal Processing

The use and application of radar technology has grown multifold in recent years [1]. Radar technology has migrated from military and defense applications to being standard components in medium to premium cars. Modern automotive radars increase safety and thus facilitate driving in both human-driven and automotive-driven cases. Slowly, radar technology has also penetrated into the industrial and consumer markets and is enabling several new applications.

The design of any radar system consists of two parts. The first part is radar hardware, including a radio frequency (RF) transceiver, waveform generator, receiver unit, antenna, and system-in-packaging. The second part is the signal processing aspect to parse the radar return echo to extract meaningful target information. Since the invention of integrated circuits, the operating frequency of transistors has been steadily increasing enabling the realization of circuit blocks that operate at frequencies up to 1 THz [2]. In parallel, transistors have been shrinking with more advanced technology nodes, allowing for further integration [3]. Figure 1.1 highlights the evolution of radar technology used in automotive applications. Silicon germanium (SiGe) bipolar technology has been the preferred silicon technology for automotive and industrial mm-wave radar over the last few years as its performance, cost, and integration level fit superbly into the application requirements [4,5]. The state-of-the-art SiGe technology has reached operating frequencies beyond 300 GHz. In [6], a FE BICMOS that can be used technology is presented with an FT of 250 GHz and a Fmax of 370 GHz for the SiGe transistors. The present technology also provides 130-nm CMOS that can be used for the realization of different radar building blocks like PLLs and DSP. RF CMOS technology also has been shown to be a candidate for mm-wave radar [7] although the RF performance is not on the same level as of

1

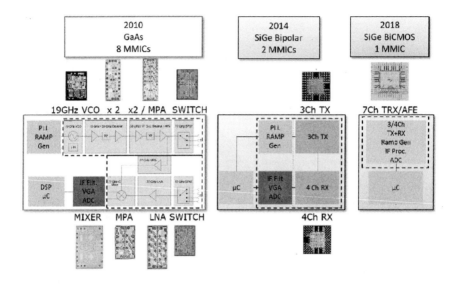

Figure 1.1   Technology trend of 77-GHz radar transceivers.

SiGe. CMOS technology offers more digital integration, which is attractive for performing signal processing on a radar chip. The high operating frequency and advanced packaging technologies have also allowed the integration of antennas into a package and in some cases into the silicon die itself. Antenna integration is essential for reducing the complexity in radar design and for reducing the overall system cost, allowing penetration of the technology into the industrial and consumer markets. It has been shown how different antenna configurations can be integrated into packaging for different field of views that cover various application requirements [8]. Going with a frequency beyond 100 GHz allows for the integration of the antenna on silicon [9], which eventually helps to reduce the cost and size of radar solutions.

   This chapter is laid out as follows: we present and introduce the different types of short-range radars in Section 1.1. In Section 1.2, we present several waveform design and ambiguity functions of radar waveforms and their properties. The system concept of a short-range radar, specifically that of a frequency modulated continuous-wave (FMCW) radar, is presented in Section 1.3. Section 1.4 presents the radar target models and their canonical structures of any radar target. In Section 1.5, three-dimensional (3D) radar data-cube processing is presented. The detection strategies clustering algorithms required for short-range radars are presented in Section 1.6, and radar target parameter estimation, namely range, velocity, and angle, are presented in Section 1.7. A detailed introduction of common tracking algorithms namely extended Kalman filters and unscented Kalman filters used in short-range radars, are presented in Section 1.8. Section 1.9 presents various applications of short-range radars in industrial, consumer, and automotive sectors.

1.1 Types of Radar

Short- and small-range radars can typically be categorized as continuous-wave (CW) radar, modulated CW radar, impulse ultrawideband (UWB) radars, and noise radars. CW radars transmit and receive waveform continuously, whereas impulse radars transmit short pulses while the receiver is not operating followed by quiet reception. Thus CW radars require distinct transmit and receive antennas and need good isolation. The major advantage of CW radars is that the signal processing at the receiver is performed at low frequency, and hence the sampling rate requirement is substantially lower compared to the transmit signal, thereby considerably simplifying the realization of processing circuitry.

1.1.1 CW Radar

CW radar transmits an unmodulated continuous frequency tone and the received echo signal is processed to estimate a target's radial velocity by evaluating the change in phase. This is caused due to a shift in Doppler frequency arising from the reflection of a moving target. However, CW radar has the disadvantage that the range information cannot be obtained. Figure 1.2 shows a block diagram of a CW radar presenting the transmit and the receive chains. However, a coarse range estimation can be obtained by either pulse-Doppler operation or transmitting two distinct frequency tones referred as frequency shift keying (FSK).

1.1.2 Modulated CW Radar

There are several modulation patterns that can modulate the transmit signal in frequency. One of the popular patterns is sawtooth frequency modulation, wherein frequency is linearly increased over time (upchirp) or decreased over time (downchirp), and thus are also referred to as linear frequency modulated (LFM). In the case of upchirp LFM, the frequency is linearly increased as

$$f(t) = f_0 + \frac{B}{T_c}t \tag{1.1}$$

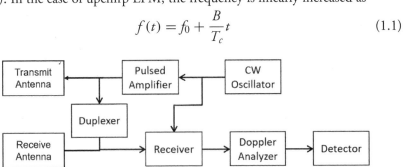

Figure 1.2 Block diagram of CW radar.

where f_0 is the ramp start frequency, B is the bandwidth, and T_c is the chirp ramp time.

The LFM waveform exhibits a near flat frequency response given as

$$S(f) = \exp\left(-j\pi \frac{T_c}{B} f^2\right) \exp\left(j\frac{\pi}{4}\right) \tag{1.2}$$

Figure 1.3 presents the upchirp LFM waveform, where the transmit (Tx) signal is an upchirp and the received (Rx) signal from two targets at different range bins. The received signal from both targets are received as a superimposition of the two at the reciever, the delay signal Δt is due to the varying range of the targets, and the shifted frequency Δf arises due to the speed of the targets. The received signal is mixed with the transmit signal followed by lowpass filtered, and this signal is referred as intermodulation frequency or IF signal.

LFM being a constant amplitude waveform is desired since it allows operating the power amplifier at saturation and thus with maximum efficiency. Further, since the chirp time and the bandwidth can be chosen independently, it offers flexibility to the radar system designer to meet different range and Doppler

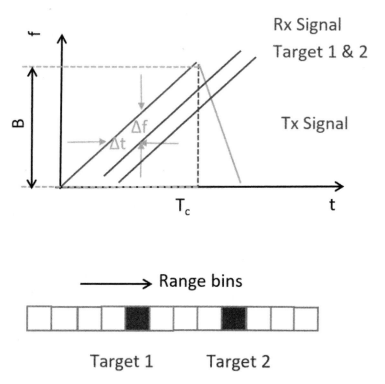

Figure 1.3   Illustration of transmit (Tx) upchirp LFM modulated signal and corresponding received (Rx) signal from two targets at different ranges.

specifications. One of the biggest advantages of frequency modulated continuous-wave radar is at the receiver the transmitted waveform is mixed with the received waveform followed by lowpass filtering, in a process called deramping. The output of the low-pass filtered intermediate frequency (IF) signal have much lower bandwidth compared to the transmitted signal that has a large sweep bandwidth, thus requiring a much lower analog-to-digital converters (ADCs) and signal processing constraints. This enables linear frequency modulated radars to be compact, low power and low-cost radar sensors.

Another commonly used waveform is triangular waveform comprising of upchirp and downchirp in sequence. Figure 1.4 presents the transmit triangular modulated waveform and the corresponding IF signal at the receiver. The beat frequency in the IF signal consists of a component due to range and Doppler as

$$f_r = \frac{B}{T_c} \cdot \frac{2R}{c}$$

$$f_d = \frac{2v}{\lambda} \tag{1.3}$$

where R, v are the range and velocity of the detected target, respectively, and λ is the carrier wavelength. The beat frequencies at the IF signal for upchirps and downchirps are given as

$$f_{bu} = f_r - f_d$$

$$f_{bd} = f_r + f_d \tag{1.4}$$

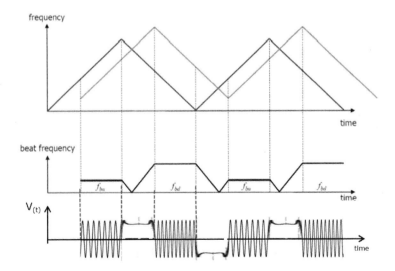

Figure 1.4   Illustration of transmit (Tx) triangular waveform with sequence of upchirp LFM and downchirp LFM and corresponding IF signal.

Thus, once the beat frequencies are determined the range and velocity of the target can be estimated as

$$R_{est} = \frac{cT_c}{4B}(f_{bd} + f_{bu})$$

$$v_{est} = \frac{\lambda}{4}(f_{bd} - f_{bu}) \tag{1.5}$$

This shows that in the range-velocity plot, the estimates could be calculated by the intersection of two lines given by (1.5). However, it is easy to see that for two targets with a different range, velocity would mean four intersecting lines, resulting in ambiguous-range velocity assignments. To solve this problem, staggered pulse repetition times (PRTs) with varying chirp time in subsequent waveform are transmitted, resulting in different slopes and facilitating unique range-velocity assignments.

In the case of a frequency modulated continuous wave, there are two critical aspects that determine the performance of such radar systems. One is the ramp linearity; any deviation from a linear ramp results in range estimation errors. The other critical aspect is the Tx-Rx leakage since the transmit and receiver are operating at the same time. The Tx-Rx leakage affects the first few range bins and also limits the radar's maximum detectable range since the dynamic range of the ADC is severally affected if it is not accounted for.

There are various frequency modulation patterns that are used in practice and in most cases require different architecture design and different processing pipelines. Apart from sawtooth and triangular frequency modulation, some of the other common frequency modulation patterns used in modern short-range radars are stepped frequency or interrupted frequency modulated waveforms. Stepped frequency modulation provides a piecewise approximation of the sawtooth linear chirp signal. The stepped frequency modulation waveform uses the same homodyne process as sawtooth frequency modulation but has much lower phase noise and has less constraints on ramp linearity. However, the stepped frequency modulation codes suffer from Doppler tolerance compared to their sawtooth equivalents. Interrupted frequency modulation tries to combine the advantage from pulse radar and frequency modulated continuous-wave radar. A linear sawtooth waveform is transmitted but to avoid the Tx-Rx leakage, the transmit signal is switched off intermittently when the receiver is receiving the echoes. However, the interrupted frequency modulated poses a strict timing constraint of the transmitter since the transmission time should be less than the round-trip propagation time of the maximum detectable target.

Apart from LFM, nonlinear frequency modulation (NLFM) can be used as transmit waveforms, which offers excellent ambiguity function properties such as lower sidelobe levels in a trade-off to complexity. Some of the nonlinear frequency modulations are

- Sine-based NLFM

$$f(t) = f_c + \frac{B}{\pi} \arcsin\left(\frac{2t}{T_c} - 1\right) \tag{1.6}$$

- Symmetrical NLFM

$$f(t) = f_c + B\left(a(t) + \sum_{n=1}^{7} K_n \sin(2\pi n a(t))\right)$$

$$\cdot \left(b(t) + \sum_{n=1}^{7} K_n \sin(2\pi n b(t))\right) \tag{1.7}$$

where $a(t) = \frac{2t}{T_c}$, $b(t) = 1 - \frac{2t}{T_c}$. $K_1 = -0.1145, K_2 = 0.0396, K_3 = -0.0202, K_4 = 0.0118, K_5 = -0.0082, K_6 = 0.0055, K_7 = -0.0040$.

1.1.3 Impulse UWB Radar

Impulse radar transmits short pulses and determines distance by measuring the time delay between the transmitted and returned signal. Impulse radar transmits ultrawideband, short-pulse waveforms. The time delay of the received pulse determines the range of the target and the peak of the spectrum determines the Doppler velocity. The pulsewidth determines the range resolution and the Doppler resolution is typically poor in such systems. However, such systems do not suffer from Tx-Rx leakage since the transmitter and the receiver do not operate at the same time.

UWB radar systems transmit signals across a much wider frequency than conventional radar systems and are usually very difficult to detect by other radio devices in their vicinity. The transmitted waveform is typically a Gaussian pulse or a derivative of a Gaussian pulse since a rectangular pulse waveform would change its shape due to the spectral limited response of the electronics circuitry apart from the target itself. The transmitted signal is significant for its very light power spectrum, which is lower than the allowed unintentional radiated emissions for electronics. The most common technique for generating a UWB signal is to transmit pulses with very short durations (less than 1 nanosecond). The spectrum of a very narrow-width pulse has a very large frequency spectrum approaching that of white noise as the pulse becomes narrower. These very short pulses need a wider receiver bandwidth compared to conventional radar systems.

1.1.4 Other Short Range Radars

Compared to frequency modulated and impulse radars, there are several other radars such as orthogonal frequency modulated (OFDM) radar, which works on the principle of a communication transceiver. In the case of an OFDM radar, the time and frequency sensitivity of OFDM waveforms are utilized to estimate the range and velocity of the target. The cyclic prefix in the OFDM transmission converts the linear convolution channel into circular convolution, thus easing the receiver processing to just a scalar division operation to estimate either the unknown communication data or unknown channel/target response. The maximum unambiguous range in the case of OFDM radars is determined by the cyclic prefix length. OFDM waveforms are typically beneficial for range-extended target responses. OFDM radars can be operated in a pure radar sensing mode, transmitting digital pseudorandom sequence as a data stream, or in a radar-communication mode; that is, the radar sensing symbols can be embedded within the time-frequency resource block with a communication data stream.

Another type of radar is noise radar, which transmits a pseudorandom signal directly from a noise-generating high-frequency source. The echo detection is based on optimal reception and correlation between the transmitted signal and the received noise waveform radar returns. Implementation of noise radar requires advanced components such as efficient noise waveform transmitters, digital correlation receivers, and wideband antennas.

Other short-range radars, such as UWB noise radar, also exists, where the transmitting waveform is a burst of noise sequence. In the case of such radar systems, transmission of the radar cannot be determined by measurement from the outside of the radar since the transmit pulses do not differ much from environmental noise, and thus are useful for interference mitigations. Table 1.1 summarizes different short-range radars and their properties [10].

Further, depending on the Tx-Rx position, a radar can be classified as monostatic if the transmit and receive antennas are located slightly apart but from the target can be viewed as approximately the same, and bistatic if the transmit and receive antennas are located at different locations.

Table 1.1
Different Short-Range Radar Waveforms and Their Key Properties

Radar Waveform	Properties
Continuous wave	Can estimate only Doppler
Frequency shift keying	Can estimate only nonzero Doppler targets, coarse range estimates
Linear FM (LFM)	Compact and lowcost due to low bandwidth IF, high peak sidelobes
Nonlinear FM (NLFM)	Tradeoff range resolution for lower sidelobes compared to LFM
UWB impulse	Ultrahigh range resolution, poor Doppler resolution
OFDM	High ADC requirements, faciliates radar-comm operations
Noise	Complex hardware circuitry, no interference effects

Irrespective of the radar type, the radar range equation provides means to the radar designer to theoretically calculate the maximum range, set transmit power, and transmit/receive antenna gain for detection of a target with a given radar cross section (RCS) through a radar link budget. Given the following radar parameters and notations:

$$\begin{cases} P_t : \text{transmit power} \\ P_r : \text{receive power} \\ G_t : \text{transmit antenna gain} \\ G_r : \text{receive antenna gain} \\ \sigma : \text{radar cross section (RCS)} \end{cases}$$

The received power at the receiver after reflection from a target with RCS σ and at range R can be expressed as

$$P_r = \frac{P_t G_t \sigma}{(4\pi R^2)^2} \tag{1.8}$$

The effective receive antenna aperture is given as $A_r = \frac{G_r \lambda^2}{4\pi}$, thus the effective received power can be expressed as

$$P_r = \frac{P_t G_t G_r \sigma \lambda^2}{(4\pi)^3 R^4} \tag{1.9}$$

Therefore, the maximum radar range can be provided as

$$R_{\max} = \left(\frac{P_t G_t G_r \sigma \lambda^2}{(4\pi)^3 P_r^{\min}} \right)^{1/4} \tag{1.10}$$

where $P_r^{\min}$ is the minimum received power required for target detection; in typical cases it is set 3 to 5 dB higher than the noise floor.

1.2 Waveform Design and Ambiguity Function

Assuming a point target, the received signal after reflection from the target can be modeled as

$$s_r(t) = \rho s(t - \tau_a) \exp j2\pi f_a t \tag{1.11}$$

where $s(t)$ is the transmit waveform and ρ is the point target radar cross section and scaling factors due to path loss and antenna gains, and so forth, (τ_a, f_a) represents the target's actual delay and Doppler components. Thus after receiver convolution processing with filter response $h(t)$, the output signal is represented as

$$y(t) = \int s_r(\tau)h(t - \tau)d\tau \qquad (1.12)$$

Therefore, the frequency spectrum of the output signal can be expressed as

$$Y(f) = \rho \int H(f)S(f - f_a)\exp(j2\pi f \tau_a)df \qquad (1.13)$$

The output SNR at time $t = \tau_a$ assuming white noise can thus be written as

$$SNR_{out}(t = \tau_a) = \frac{\rho^2 \left| \int H(f)S(f)df \right|^2}{N_0 \int |H(f)|^2 df} \qquad (1.14)$$

where N_0 is the noise power.

By Cauchy-Schwarz inequality, the numerator follows the following inequality

$$\left| \int H(f)S(f)df \right|^2 \leq \int |H(f)|^2 df \int |S(f)|^2 df \qquad (1.15)$$

where the equality follows when $H(f) = S^*(f)$, which means the optimal receiver filter processing for maximizing the output SNR corresponding to a point target with an assumption of white noise is a matched filter, where the received signal $s_r(t)$ is mixed with the transmit waveform followed by lowpass filtering (representing the integration function in (1.12)). The signal power at $t = \tau_a$ is equal to $|\int |S(f)|^2 df|^2$ and for a fixed frequency bandwidth this is maximized by a flat transmit frequency spectra, such as LFM waveform.

However it must be noted that in the case of an extended target or colored noise interference, the conventional matched filter is no longer the optimal receiver processing, and the flat frequency spectra transmit waveform is also not the optimal waveform in terms of output detection SNR.

The pulse compression gain [11] denotes the ratio of signal-to-noise ratio (SNR) improvement achieved by the receiver filter processing and is defined as

$$G = \frac{SNR_{out}(t = \tau_a)}{SNR_{in}} \qquad (1.16)$$

where input SNR, SNR_{in} can be written as

$$SNR_{in} = \frac{\rho^2(1/T_c) \int |S(f)|^2 df}{N_0 B} \qquad (1.17)$$

On substituting SNR_{out} and SNR_{in}, (1.16) for compression gain can be simplified to

$$G = T_c B = \frac{T_c}{\tau_B} \qquad (1.18)$$

where τ_B denotes the compressed pulsewidth at the output of the filter matched to the transmit waveform. This shows the effect of the matched filtering as a compression gain, which is equal to the time-bandwidth product. Thus the output SNR can be increased by either increasing the time or the bandwidth of the transmit waveform, while retaining the transmit power. This is an important design consideration for the radar system designer since the transmit power is typically limited by regulations to avoid interferences to other radio devices.

The other important aspect in radar signal processing is the Woodward ambiguity function, $\chi(\tau, \nu)$, which is a two-dimensional time delay and Doppler frequency response of the received radar signal from a point target due to a receiver matched filter. The Woodward or self-ambiguity function is determined by the properties of the transmitted waveform irrespective of interference, noise, or target scenario (multitarget or spread targets). The Woodward ambiguity function of any waveform can be defined as

$$\chi(\tau, \nu) = \int_{-\infty}^{\infty} s(t)s^*(t + \tau)e^{j2\pi \nu t}dt \qquad (1.19)$$

where $s(t)$ is the transmitted waveform, and (τ, ν) are the mismatched delay and Doppler shifts.

In the case of basic FMCW waveform, the Woodward ambiguity function has a closed-form expression as follows

$$\chi(\tau, \nu) = \left(1 - \frac{\tau}{T_p}\right) sinc\left(\pi T_p\left(\nu \mp B\left(\frac{\tau}{T_p}\right)\right)\left(1 - \frac{\tau}{T_p}\right)\right) \qquad (1.20)$$

where $sinc(x) = sin(x)/x$. Figure 1.5 presents the 3D woodward ambiguity function and the corresponding 2D contour of the ambiguity function for an LFM waveform. The observed mainlobe is actually part of a delay-Doppler ridge that exhibits a gradual roll-off from the peak at (0,0). The existence of this ridge is why LFM is also referred to as a Doppler-tolerant waveform, since an appreciable Doppler shift induces little SNR loss relative to the peak.

The Woodward ambiguity function of a rectangular pulse waveform, whose pulse width is T_c can be expressed as

$$\chi(\tau, \nu) = \left|\left(1 - \frac{|\tau|}{T_c}\right) sinc\left(\pi \nu T_c\left(1 - \frac{|\tau|}{T_c}\right)\right)\right| \qquad (1.21)$$

In the case of a pulse train of FMCW waveform, the ambiguity function is given as

$$\chi^{fmcw}(\tau, \nu) = \left| \int_{-\infty}^{\infty} \sum_{m=1}^{M} \sum_{m'=1}^{M} s(t - mT_p)s(t - m'T_p - \tau) \right.$$

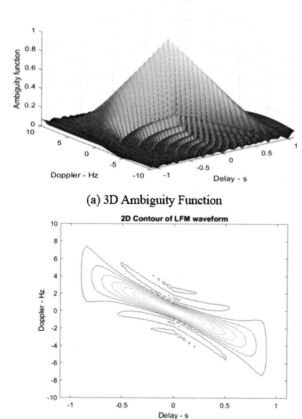

(a) 3D Ambiguity Function

(b) 2D Contour Ambiguity Function

Figure 1.5   (a) 3D Woodward ambiguity function of LFM waveform, (b) 2D contour of Woodward ambiguity function of LFM waveform.

$$\left. P(t - mT_p)P(t - \tau - m'T_p)e^{i2\pi vt}dt \right|$$

where M is the train of consecutive chirps transmitted by the radar sensor. The Woodward ambiguity function has several properties, such as

1. Maximum at (0,0); that is,

$$\chi(\tau, v) \leq \chi(0,0) = 1 \quad \forall(\tau, v)$$

2. Constant volume; that is,

$$\int_{-\infty}^{\infty} \int_{-\infty}^{\infty} \chi(\tau, v)^2 d\tau dv = 1$$

3. Symmetric at origin; that is,

$$\chi(-\tau, -\nu) = \chi(\tau, \nu)$$

4. Linear FM effect; that is,

$$u(t) \implies \chi_u(\tau, \nu), \quad \text{then} \quad u(t)e^{j\pi kt^2} \implies \chi_u(\tau, \nu - k\tau)$$

The Woodward ambiguity function has different interpretations among radar practitioners. One school of thought is that the ideal ambiguity function is a thumback delta dirac function, and is expressed as $\chi(\tau, \nu) = \delta(\tau, \nu)$ which means the matched filter would respond to a correct estimate of range and Doppler. The search for an ideal thumback response for the ambiguity function has lead to several waveform research such as Barker code, Golay complementary pairs, Frank Zadoff Chu codes, and Gold sequence. The other school of thought is that such response is not usually desirable since a nonzero Doppler would not remove the target response along the range transform. Thus a range-Doppler coupling effect, as exhibited by LFM waveform, is preferred. As a result, the ambiguity function presents an important tool to study the waveform design and properties suitable for radar sensing functions.

Apart from the narrowband Woodward ambiguity function, there are several variants that exist, such as the wideband ambiguity function, spread ambiguity function, pulse train ambiguity function, and expected ambiguity function. The general cross ambiguity function can be defined as

$$\chi_{\text{cross}}(\tau_h, f_h; \tau_a, f_a) = \left| \int s(t; \tau_a, f_a) \, g^*(t; \tau_h, f_h) \, dt \right|$$

$$= \left| \int S(f; \tau_a, f_a) \, G^*(f; \tau_h, f_h) \, df \right| \qquad (1.22)$$

for the cross ambiguity function, where (τ_h, f_h) are the hypothesized parameters in the receiver and (τ_a, f_a) are the true range and Doppler parameters. $s(t), S(f)$ are the time domain and frequency response of the transmitted waveform, $g(t), G(f)$ are the time and frequency response of the mismatched signal. The above equation can be extended to derive a closed-form expression for a cross ambiguity function in presence of a point target. It can also be extended to derive the ambiguity function in the case of the colored-noise matched filter where receiver filter $g(t)$ is not matched to the transmit waveform. This will be a case when the noise is not white, thus requiring a prewhitening operation provided by the colored-noise matched filter.

1.3 System Concept

In this section, we describe the system concept of a typical short-range FMCW radar. Figure 1.6 presents the block diagram of a typical short-range FMCW radar

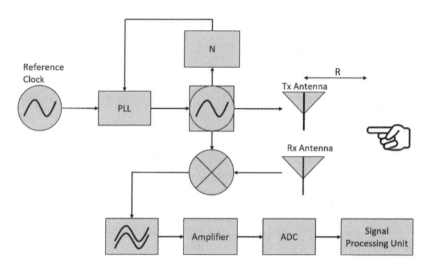

Figure 1.6 Functional block diagram of short range FMCW radar.

with 1 transmit, 1 receive antenna. The radar features power amplifiers, low-noise amplifiers, voltage-controlled oscillators, and analog-to-digital converters (ADCs). The voltage-controlled oscillator module generates a linear frequency-modulated continuous-wave signal, which is referenced through a local oscillator. The generated LFM waveform is then amplified by the power amplifier before being transmitted from the antenna. The receive antenna receives the target-distorted transmit waveform, which is then amplified by the low-noise amplifier. This receiver amplified RF signal is then mixed with the transmitted signal followed by a lowpass filter to produce the beat intermediate frequency. The intermediate frequency signal is then passed to the ADC for subsequent processing. As mentioned earlier, since the intermediate frequency has low bandwidth, the requirements for ADC and signal processing are drastically lower compared to sampling the transmit waveform with large bandwidths, thus enabling FMCW radars to be compact, small, and low-cost solutions.

Figure 1.7(a) depicts the de-ramping operation at the receiver. $s_{TX}(t)$ and $s_{RX}(t)$ refers to the transmit and corresponding received chirp, respectively. The round-trip propagation delay $\tau = 2R/c$ gets translated to intermediate frequency after mixing at the receiver. And thus spectral analysis along the chirp provides the range estimation of the targets in the radar's field of view. The swept bandwidth B determines the range resolution as $\delta R = c/2B$. The maximum unambiguous range turns out to be $R_{max} = N_s \delta R$, where N_s is the number of transmit frequency steps. The ADC output along a single chirp is referred to as fast time in the literature.

Figure 1.7(b) shows the frame structure of FMCW radar. The chirp duration T_c determines the maximum detectable unambiguous Doppler

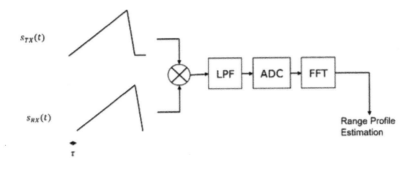

$s_{TX}(t)$

$s_{RX}(t)$

τ

LPF → ADC → FFT

Range Profile
Estimation

(a) Deramping process

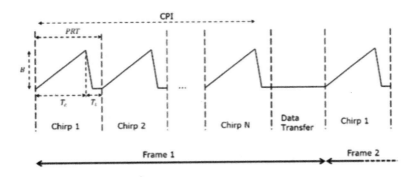

CPI

PRT

B

T_c T_i

Chirp 1   Chirp 2   Chirp N   Data Transfer   Chirp 1

Frame 1   Frame 2

(b) Frame Structure

Figure 1.7 (a) Deramping processing of FMCW signal, and (b) frame structure of FMCW radar.

$f_{d_{max}} = 1/2(T_c + T_i)$, whereas T_i denotes ignored time duration, which represents the time for resetting the start frequency, DAC, and ADC sampling. The time duration between two chirps is referred to the PRT or pulse repetition interval (PRI) and is given as $PRT = T_c + T_i$. The velocity content of the target at a range bin causes phase change across multiple chirps at the target's range bin. Thus the spectral estimation across chirps provides the velocity information of the target. The collection of consecutive chirps used for coherent integration is referred to as the frame or dwell, and represents how often target parameters are estimated or updated. The coherent processing interval (CPI), which is the time duration in a frame, determines the Doppler resolution $\delta f_d = 1/2CPI$. The time samples along the chirps within the coherent processing interval is referred to as slow time in the literature.

Correct raw data acquisition as the first step is ensured by setting the start time and sampling frequency of the ADC so that the required N_S number of DAC/ADC samples are equally distributed within the ramp start and end. For each frame having N_{RX} number of Rx channel, a three-dimensional data cube $\Phi \in \mathcal{C}^{N_S \times N_C \times N_{RX}}$ containing the complex-valued baseband signals is obtained. The first dimension contains all samples per chirp (fast-time) for range estimation, the second dimension belongs to the different chirps per frame (slow-time) for velocity estimation, and the third dimension corresponds to the N_{RX} receive antennas for angle of arrival (AoA) estimation.

In the case of multiple-input, multiple-output (MIMO) radar, it is important to maintain orthogonality at the receive among transmit waveforms to ensure independent data channel streams. By enabling MIMO configuration, the virtual channels can be computed as a 2D convolution of the transmit spatial and recieve spatial locations thus increasing the angular resolution and probability of detection. The orthogonal transmit waveform can be achieved by distributing chirps across time, frequency, and codes. Figure 1.8 depicts the time division multiplexing (TDM), frequency division multiplexing (FDM) and slow-time code division multiplexing (CDM) scheme [12]. Time-division multiplexed MIMO radar is the simplest configuration and is achieved by transmitting the same chirp in a round-robin fashion across all transmit antennas. However, this simple configuration has the maximum unambiguous Doppler limited by the inverse of the number of transmit antennas. In the frequency division multiplexing configuration, each transmit antenna transmits a different part of the frequency sweep. This configuration requires different voltage-controlled oscillators for each transmit channel and although it preserves the maximum unambiguous Doppler it suffers a loss of range resolution for each transmit waveform. On the other hand, slow-time CDM makes use of all the time-frequency resources, thus retaining the maximum unambiguous Doppler and also the range resolution of a single-input, single-output (SISO) FMCW radar, while at the same time increasing the number of virtual antennas leading to enhanced angular resolutions and also increasing the probability of detection.

1.4 Target Model

For a long-range radar with a narrow frequency sweep, targets appear as a point target. However, with a wide frequency sweep targets appear as a range-spread target instead of a point target at the radar receiver. Further radar responses from targets such as humans are in general spread across Doppler as well due to the macro-Doppler component of the torso and associated micro-Doppler components due to hand, shoulder, and leg movements. Thus, human targets are perceived as doubly spread targets across range and Doppler. The simplest extended target models are based on the weak scattering model, which utilizes

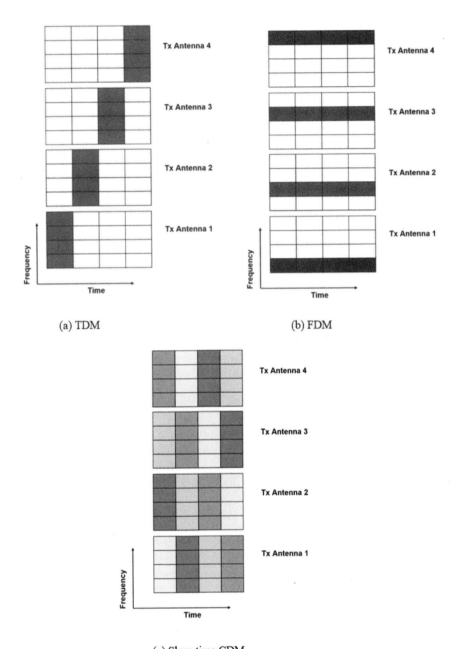

(a) TDM

(b) FDM

(c) Slow-time CDM

Figure 1.8   Chirp distribution across time, frequency, and transmit antennas: (a) time division multiplexing, (b) frequency division multiplexing, and (c) slow-time code division mutliplexing.

the idealized concept of persistent, localized scattering centers. The scattering centers in this model are the outcome of the geometrical approximation and are represented as

$$h(\tau, v, \theta) = \sum_{n=1}^{N} a_n \delta(\tau - \tau_n) \delta(v - v_n) \delta(\theta - \theta_n) \qquad (1.23)$$

where $h(\tau, v, \theta)$ is the target impulse response as a function of delay, Doppler and azimuth angle of each scattering center represented by the parameters $a_n, \tau_n, v_n, \theta_n$ denote the radar cross section amplitude, delay, Doppler, and angle of the nth scattering center. The target impulse response is characterized by the superposition of N such scattering centers. The weak scattering model has limitations and is ill-behaved for scatterers that include edges, gentle curved plates, and reentrant structures such as ducts and cavities. Further, the model does not capture variations in the target impulse response arising from aspect angle changes and other nongeometric processes. However, the weak scatterer model can be used to understand the target response due to its geometric properties.

The radar views the space in polar coordinates, thus the 3D Cartesian position of the kth target and its corresponding unit vector can be represented as

$$r_k = \begin{bmatrix} R_k \cos(\theta_k) \sin(\phi_k) & R_k \cos(\theta_k) \cos(\phi_k) & R_k \sin(\theta_k) \end{bmatrix}$$
$$u_k = \begin{bmatrix} \cos(\theta_k) \sin(\phi_k) & \cos(\theta_k) \cos(\phi_k) & \sin(\theta_k) \end{bmatrix} \qquad (1.24)$$

where θ_k and ϕ_k are the elevation and azimuth angle, respectively, of the target with respect to the center of the virtual radar array. For estimation of the target in a 3D space, a MIMO configuration of at least $N_{TX} = 2$ transmit elements and $N_{RX} = 2$ receive elements in an L-shaped linear array fashion with appropriate spacing is required. This results in a virtual 2×2 rectangular array configuration sufficient for estimation of target's elevation and azimuth coordinates. As depicted in Figure 1.9, the 3D positional coordinates of the TX element are denoted as d_m^{TX}, $m = 1, \ldots, N_{TX}$ and the RX element as d_n^{RX}, $n = 1, \ldots, N_{RX}$ in space. On assuming far-field conditions, the signal propagation from the TX element d_m^{TX} to a point scatterer p and subsequently the reflection from p to RX element d_n^{RX} can be approximated as $2R_k + d_{mn}$, where R_k is the base distance of the kth scatterer to the center of the virtual linear array and d_{mn} refers to the relative position of the virtual element to the center of the array. Figure 1.9 presents the 3D position of the target with reference to mth transmit and nth receive antennas.

The transmit steering vector can be written as

$$a_m^{TX}(\theta, \phi) = \exp\left(-j2\pi \frac{d_m^{TX} u(\theta, \phi)}{\lambda}\right); \quad m = 1, \ldots, N_{TX} \qquad (1.25)$$

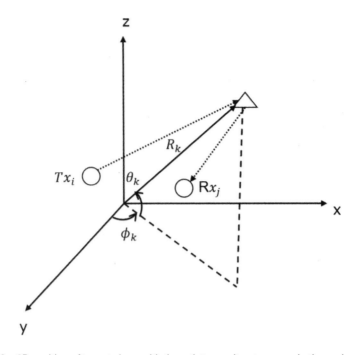

Figure 1.9 3D position of target along with the *m*th transmit antenna and *n*th receive antenna.

while the receiving steering vector is

$$a_n^{RX}(\theta,\phi) = \exp\left(-j2\pi\frac{d_n^{RX}u(\theta,\phi)}{\lambda}\right); \; n = 1,\dots,N_{RX} \qquad (1.26)$$

where λ is the wavelength of the transmit signal.

The received baseband signal from the kth target scatterer can be expressed as

$$\bar{s}_{RX}(t) = \rho_k e^{\frac{-j2\pi 2u_k \cdot r_k}{\lambda}} a_n^{Rx}(\theta_k,\phi_k) a_m^{Tx}(\theta_k,\phi_k)^T \bar{s}_{TX}(t) \qquad (1.27)$$

where ρ_k represents the composite amplitude contribution due to propagation path loss, antenna gains, and receiver gains. $\bar{s}_{TX}(t)$ and $\bar{s}_{RX}(t)$ are the transmitted signals from N_{TX} transmit antennas and the received signal at N_{RX} receive antennas, respectively. After having estimated the range R_k of the kth target through spectral analysis along fast time, the angular coordinates of the target, namely azimuth angle θ_k and elevation angle ϕ_k, can be estimated through monopulse, Capon, or FFT beamforming algorithms.

Compared to the weak scattering center model, there are several rigorous models that take the physical characteristics of the scattering center into consideration. Some of such approximate target models are geometric theory of diffraction (GTD) [13,14] and uniform theory of diffraction (UTD). Figure 1.10

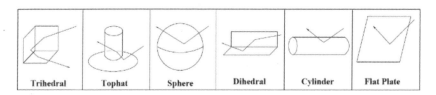

Figure 1.10 GTD primitive scattering structures.

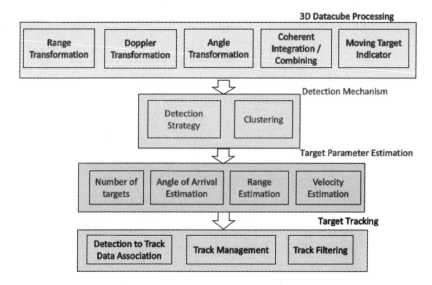

Figure 1.11 Overall standard radar processing flow.

presents the canonical structures, such as flat plate, top hat, trihedral corner reflector, dihedral corner reflector, cylinder, or sphere, which comprises the target response. The RCS response for a sphere is 1, the RCS response of some of the structures can be expressed as

$$S(k, \phi, \theta) = (jk)^\alpha sinc(kL_n \sin(\phi) \cos(\theta)) \tag{1.28}$$

where k is the wavenumber, $L_n = 0$ if the scattering center is localized and $L_n \neq 0$ if distributed. The parameter α has a half-integer value.

1.5 3D Data-cube Processing

Figure 1.11 presents the overall radar signal processing blocks involving 3D data-cube processing, detection mechanism, target parameter estimation and target tracking. In this section, we present the 3D data-cube processing. The transmitted LFM on being reflected from the target is mixed with a replica of the transmitted signal resulting in a beat signal. The phase of the beat signal after the mixer due to the kth point target can be expressed as

$$\phi_k(t) = 2\pi \left(f_c \tau_k + \frac{B}{T_c} t \tau_k - \frac{B}{2T_c} \tau_k^2 \right) \tag{1.29}$$

where $\tau_k = \frac{2(R_k + v_k t)}{c}$ is the round-trip propagation delay between the transmitted and received signal after reflection from the kth target with range R_k and radial velocity v_k. The downconverted IF signal therefore is the superposition of received signal from K point scatterers and thus expressed as

$$s_{IF}(t) = \sum_{k=1}^{K} \exp \left(2\pi \left(\frac{2f_c R_k}{c} + \left(\frac{2f_c v_k}{c} + \frac{2BR_k}{cT_c} \right) t \right) \right) \tag{1.30}$$

after ignoring the second-order terms $\frac{2Bv_k}{T_c c} t^2$. The frequency shifts due to range and velocity arising from multiple point targets at the IF signal are decoupled by generating range-Doppler images (RDI) across all virtual channels. Expanding the time index t as $n_k T_{\text{frame}} + n_s T_{\text{PRT}} + n_f$, where n_f is the fast time index $0 < n_f < T_c$, and n_s denotes the slow time index, T_{PRT} is the chirp repetition time indicating the time difference between start of two consecutive chirps in a frame and n_k denotes the frame number.

Figure 1.12 presents the 3D radar data cube that depicts data across the fast time, slow time, and virtual channels. Now based on the processing complexity and the specific application, the 3D data cube can be processed through a 1D transformation, followed by detection of the targets in that domain and then

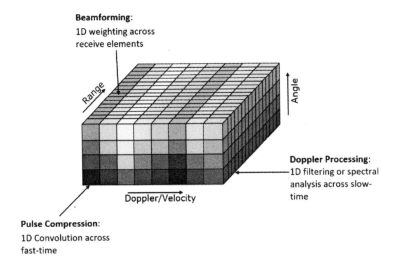

Figure 1.12 3D radar data cube.

estimation of the other target parameters. Alternately the target is detected in one of the 2D slices and the other target parameter is estimated.

1.5.1 1D Processing

Figure 1.13(a) depicts the range processing step, wherein the fast time data is transformed into range bins through a 1D FFT. The fast-time data is first applied with a windowing function and then optionally zero-padded to increase the range accuracy as

$$R_i^n(k) = \sum_{l=0}^{N_S-1} r(l)w(l)\exp\left(-j\frac{2\pi ln}{Z}\right), \quad 0 \le n < Z \tag{1.31}$$

where Z is the zero pad length, N_S is the number of ADC samples per chirp, and $R_i^n(k)$ is the range bin value at ith chirp and nth range bin on kth frame. $w(l)$ is the window function and is applied to reduce the sidelobe level from -13 dB (no windowing) to a much lower acceptable value. However, the window function exhibits a trade-off of sidelobe levels with the main-lobe width. Thus the theoretical range resolution, $\delta r = \frac{c}{2B}$, is increased by a some factor, which is a function of the windowing function used. Some of the standard window functions

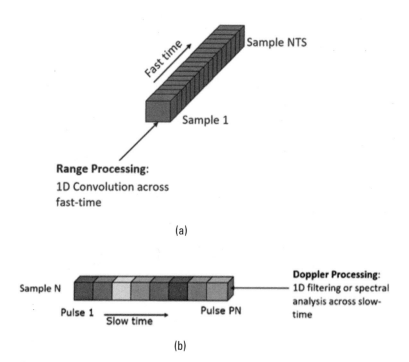

Figure 1.13   (a) Range processing data, and (b) Doppler processing data.

are Hanning window, Blackman-Harris window, Kaiser window, raised-root cosine window function, and Chebysev window function. Note that in (1.31) the summation runs up to N_S instead of Z since beyond N_S the samples are zeros. Further it must be noted that to ensure proper scaling the amplitude spectra should be scaled by $\frac{1}{N_S}$ instead of $\frac{1}{Z}$ for the same reason. The 1D range bins depicts the range position of all the targets in front of the sensor. Once the target range bins are detected, the Doppler and angle of the targets can be estimated with Doppler FFT and mono-pulse algorithm of angle FFT, respectively. Similar preprocessing steps are applied across slow-time and virtual channels if FFT transformation are used.

Following the range transformation, pulse integration is applied to combine the data over all chirps to improve the signal-to-noise ratio (SNR) facilitating subsequent target detection. Coherent integration combines the phase and magnitude of the range FFT data coherently over all chirps of a frame, and is applied at frame k. $R_{ci}(k)$ is based on the mean of the range FFT data $\{R_i^n(k)\}_{i=1}^{N_C}$ over N_C chirps:

$$R_{ci}^n(k) = \frac{1}{N_C} \cdot \sum_{i=1}^{N_C} R_i^n(k) \tag{1.32}$$

In the case of additive white Gaussian noise, owing to the coherent integration, the noise power is reduced by a factor of N_C. Considering noise $z_i^n \sim \mathcal{N}(0, \sigma_z^2)$ to be independent and identically distributed (I.I.D.) then $\frac{1}{N_c} \sum_{i=1}^{N_c} z_i^n \sim \mathcal{N}(0, \frac{1}{N_c} \sigma_z^2)$, thus improving the signal-to-noise ratio.

In the case of noncoherent integration, the integration is applied just over the range spectra, which results in a fractional signal-to-noise ratio improvement compared to their coherent integration counterparts. The coherently integrated range spectrum contains reflections from not only the moving objects but also the stationary objects in its field of view. If the application is specific to detecting moving targets such as humans, moving target indicator (MTI) filter is used. The MTI filter can be used to suppress the contribution of these stationary objects and also removes the static Tx-Rx leakage. One of the means of applying the MTI filter is through a moving average filter so that at each frame the coherent integrated range FFT spectrum, denoted as $R_{ci}(k)$ at kth frame, can be expressed as

$$R_{ci}(k) = R_{ci}(k) - S(k-1)$$
$$S(k) = \alpha \cdot R_{ci}(k) + (1-\alpha) \cdot S(k-1) \tag{1.33}$$

where α is the forget factor, for example set to 0.01. This filtered range FFT spectrum can then be utilized for the subsequent target detection.

Figure 1.13(b) depicts the Doppler processing step, and for the cases where the desired signal component across Doppler is modeled as a discrete sinusoid at frequency ν, the signal across the slow time can be represented as

$$s = \sigma \left[1 \quad e^{j2\pi \nu T_{\text{PRT}}} \quad e^{j2\pi \nu 2 T_{\text{PRT}}} \quad \cdots \quad e^{j2\pi \nu N_c T_{\text{PRT}}}\right] \qquad (1.34)$$

Thus the matched filter output in this case of discrete sinusoid in Doppler amounts to a discrete Fourier transform (DFT); that is,

$$h = \sigma \left[1 \quad e^{-j2\pi \nu T_{\text{PRT}}} \quad e^{-j2\pi \nu 2 T_{\text{PRT}}} \quad \cdots \quad e^{-j2\pi \nu N_c T_{\text{PRT}}}\right] \qquad (1.35)$$

However, in the case of human sensing where the Doppler component is not just the macro-Doppler or single sinusoid but is modeled as a superposition of a micro-Doppler signature along with the macro-Doppler component, the DFT is no longer the matched filter operation that maximizes the output SNR. In the case of FSK or pulsed-CW radars, the Doppler transform is first applied, followed by range estimation.

In the case of Doppler processing when time division multiplexing of chirps in a MIMO configuration is used the maximum unambiguous Doppler is reduced by a factor of the number of transmit antennas. In such cases, the unambiguous Doppler can be increased by staggered PRT concept. Say $\text{PRF}_1, \text{PRF}_2, \cdots \text{PRF}_T$ are the pulse repetition frequency of group of chirps transmitted in a frame, where PRF_t are chosen so that they are integers and relative multiples of some frequency (i.e., $\text{PRF}_t = p_t \text{PRF}_0$). By application of the Chinese remainder theorem, the maximum unambiguous Doppler turns out to be $f_{\text{max}} = p_1 p_2 \cdots p_T \text{PRF}_0$, thus increasing the unambiguous Doppler frequency. Further, if the pulse repetition time is increased it leads to a lower duty cycle, and therefore lower power consumption, and is an effective processing technique in power-critical applications.

As depicted in Figure 1.14, the phase difference of the received signal between two antennas is provided as $\delta\phi = 2\pi \frac{d}{\lambda} \sin(\theta)$, where d is the distance between the two antennas, λ is the wavelength of the receive waveform, and θ is the angle extended from the boresight with respect to the receive wavefront. In the case where there is only one target at a range bin, the phase monopulse algorithm is the maximum likelihood estimator for estimating the angle of arrival and can be expressed as

$$\hat{\theta} = \sin^{-1}\left(\frac{\lambda \hat{\delta\phi}}{2\pi d}\right) \qquad (1.36)$$

where $\hat{\delta\phi}$ is the estimated phase difference between received data on both antennas. It is easily noted that the system becomes underdetermined if there is more than one target at the range bin with two antennas. Thus, for N_t, multiple-target angle of arrival estimation requires at least $N_t + 1$ virtual receive antennas to form an overdetermined system of equations.

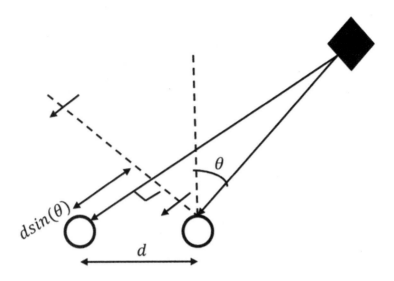

Figure 1.14   Angle of arrival.

Alternately, 2D transformation are applied before detection and estimation of parameters. Based on the application, either RDI, range-cross range images (RCRIs), or angle-Doppler images are generated. The following sections describe the generation of these images before further processing.

1.5.2   2D Range Doppler Images

The received signal at frame n_k, $s_{IF}(t; n_k)$, from consecutive chirps are arranged in the form of a 2D matrix; that is, $s_{IF}(n_s, n_f; n_k)$. The RDI is generated for each channel by applying a window function, zero-padding, and then a 1D fast Fourier transform (FFT) along fast time to obtain the range transformation, followed by applying window function, zero-padding and then 1D FFT along slow time index. The 2D discrete Fourier transform (DFT) transforms the sampled signal $s_{IF}(n_s, n_f; n_k)$ into range-Doppler domain

$$S(p, q, n_k) = \sum_{n_s=1}^{Z_{Nc}} \left(\sum_{n_f=1}^{Z_{NTS}} w_f(n_f) s_{IF}(n_s, n_f; n_k) e^{-j2\pi p n_f / Z_{NTS}} \right)$$
$$\cdot w_s(n_s) e^{-j2\pi q n_s / Z_{Nc}} \tag{1.37}$$

with NTS and Z_{NTS} being the number of transmitted samples defined by digital to analog converter (DAC) sampling points over chirp duration and zero-padding along fast-time, respectively. Nc and Z_{Nc} are being the number of chirps in a frame and zero-padding along the slow-time, respectively. $w_f(n_f)$ and

$w_s(n_s)$ represent the window function along fast-time and slow-time, respectively, and for our implementation we have used the Hamming window and Kaiser window, respectively. p, q denotes the index over range and Doppler dimensions, respectively. It is obvious that the peaks in the range-Doppler domain occur at

$$p_k = \left(\frac{2f_c}{c} v_k + \frac{2B}{cT_c} R_k \right)$$

$$q_k = \frac{2v_k f_c}{c} \tag{1.38}$$

Using sawtooth FMCW with fast ramps, $\frac{2f_c}{c} v_k \ll \frac{2B}{cT_c} R_k$, the range peaks appear at $\frac{2B}{cT_c} R_k$. The maximum velocity is given as

$$v_{\max} = \frac{c}{2f_c T_{\mathrm{PRT}}}$$

and the minimum velocity is

$$\delta v = \frac{c}{2f_c Z_{Nc} T_{\mathrm{PRT}}}$$

Figure 1.15 shows the signal processing steps to create the processed RDIs. Following the 2D FFT to transform the data into range-Doppler domain, background subtraction is achieved through a moving average filter as

$$S(p, q; n_k) = S(p, q; n_k) - S_B(p, q; n_k)$$

$$S_B(p, q; n_k + 1) = \gamma S_B(p, q; n_k) + (1 - \gamma) S(p, q; n_k) \tag{1.39}$$

where $S(p, q; n_k)$ is the RDI at n_k^{th} frame, and $S_B(p, q; n_k)$ is the background RDI at the n_k^{th} frame, and γ is the moving average coefficient.

Following the RDI generation across all virtual channels, to gain diversity and improve signal quality, maximal ratio combining is used to combine the RDIs from different antennas. The gains for the weighted averaging are determined by estimating the SNR for each RDI across antennas. The effective RDI is computed as

$$\mathrm{RDI}_{\mathit{eff}} = \frac{\sum_{rx=1}^{N_{\mathrm{Rx}}} g^{rx} |\mathrm{RDI}^{rx}|}{\sum_{rx=1}^{N_{\mathrm{Rx}}} g^{rx}} \tag{1.40}$$

where RDI^{rx} is the complex RDI of the rxth receive channel, and the gain is adaptively calculated as

$$g^{rx} = (\mathrm{NTS.PN}) \frac{\max\{|\mathrm{RDI}^{rx}|^2\}}{\left(\sum_{l=1}^{\mathrm{NTS}} \sum_{m=1}^{\mathrm{PN}} |\mathrm{RDI}^{rx}(m, l)|^2 - \max\{|\mathrm{RDI}^{rx}|^2\} \right)} \tag{1.41}$$

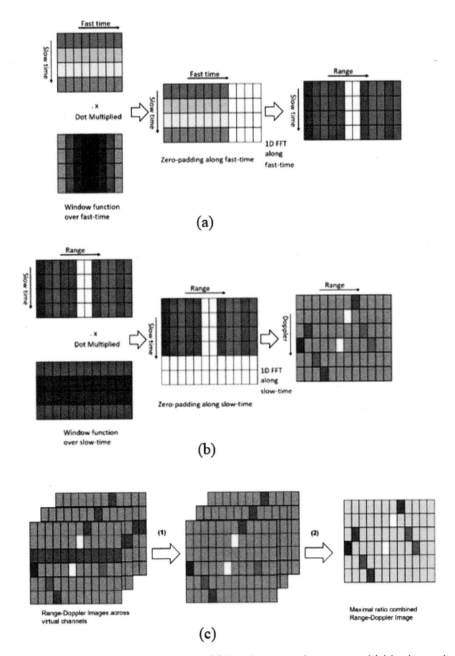

Figure 1.15 (a) Range processing steps, (b) Doppler processing steps, and (c) background subtraction and combining.

where max{.} represents the maximum value from the 2D function, and g^{rx} represents the estimated SNR at rth receive channel. Thus the ($PN \times NTS \times N_{Rx}$) RDI tensor is transformed into a ($PN \times NTS$) RDI matrix, which is then subsequently fed to the detection algorithm.

1.5.3 Range Cross-Range Images

In this book, range cross-range images (RCRI) and range-angle images (RAI) are used interchangeably, since one can simply be obtained from the other through coordinate transformation. Figure 1.16 presents the algorithm flow of how the range transform across the virtual channels are transformed into range-angle across all the Doppler indices, which is further combined to form a single range-angle image used for detection and further processing. There are several approaches to obtain the range-angle image. The simplest approach takes a FFT along the virtual channel after applying a window function and zero-padding.

The $N_t \times N_r$ deramped beat signal can be stacked into a vector and the Kronecker product of the steering vector of the Tx array $a^{Tx}(\theta)$ and the steering vector of the Rx array $a^{Rx}(\theta)$; that is, $a^{Tx}(\theta) \otimes a^{Rx}(\theta)$ can be used to resolve the relative angle θ of the scatterer. Subsequently, beamforming of the MIMO array signals can be regarded as synthesizing the received signals with the Tx and Rx steering vectors. The azimuth imaging profile for a range bin l can be generated using the Capon spectrum from the beamformer. The Capon beamformer is computed by minimizing the variance/power of noise while maintaining a distortionless response toward a desired angle. The corresponding

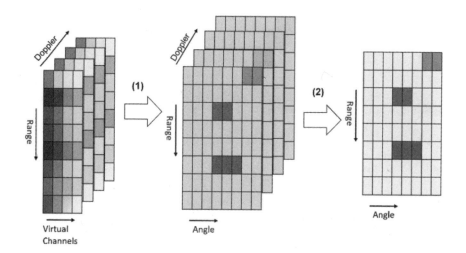

Figure 1.16 Range-angle transformation followed by maximal ratio combining.

quadratic optimization problem is

$$\min_w w^H C w$$

$$s.t. \quad w^H \left(a^{Tx}(\theta) \otimes a^{Rx}(\theta) \right) = 1 \tag{1.42}$$

where C is the covariance matrix of noise, the above optimization has a closed-form expression given as $w_{\text{capon}} = \frac{C^{-1} a(\theta)}{a^H(\theta) C^{-1} a(\theta)}$, with θ being a desired angle. On substituting w_{capon} in objective function of (1.42), the spatial spectrum is given as

$$P_l(\theta) = \frac{1}{\left(a^{Tx}(\theta) \otimes a^{Rx}(\theta) \right)^H \hat{C}_l^{-1} \left(a^{Tx}(\theta) \otimes a^{Rx}(\theta) \right)} \tag{1.43}$$

$$\text{with } l = 0, ..., L$$

However, estimation of noise covariance at each range bin l is difficult in practice, and hence $\hat{C}_l$ is estimated, which contains the signal component as well and can be estimated using the sample matrix inversion (SMI) technique $\hat{C}_l = \frac{1}{N} \sum_{k=1}^{K} s_l^{IF}(k) s_l^{IF}(k)^H$, where K denotes the number of snapshots used for the signal-plus-noise covariance estimation and $s_l^{IF}(k)$ is the deramped intermediate frequency signal at range bin l with k being the frame index.

The Capon spatial spectrum can be viewed as an adaptive signal-dependent spatial filtered spectrum that is characteristic or defined by the material the radar is illuminating. Capon spectra capture the spatial contour of the material intensity reflection and thus are more robust to increase in noise power, antenna array errors, and target aspect angle.

Alternately the range-cross range image can be generated by using simple beamforming algorithm based on phase monopulse algorithm. The range-angle images can be expressed as

$$P_l(\theta) = \sum_{i=1}^{N_v} \left(a^{Tx}(\theta) \otimes a^{Rx}(\theta) \right) s_l^{IF}(k) \tag{1.44}$$

Figure 1.17 presents the 2D slices of the 3D radar data cube that can be processed for either RDI, RCRI, or angle-Doppler images based on different applications for subsequent processing.

1.6 Detection Strategy and Clustering

1.6.1 Detection Algorithm

Following either the 1D range transformation or 2D range-Doppler/range cross-range transformation, detection strategy is applied to detect if a cell under test

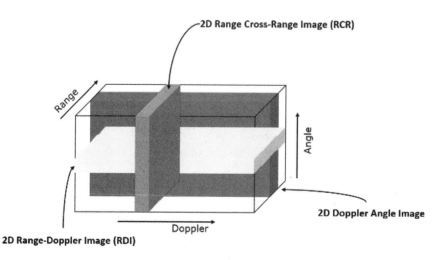

Figure 1.17   3D data-cube processing with 2D slices.

(CUT) contains a target. The detection strategy can be applied over the 3D data cube; however, it is seldom done on an embedded processor mainly due to its computational complexity. Figure 1.18(a) presents the detection strategy applied to a 1D data vector, which can be the range transformation or the Doppler transformation data. Figure 1.18(b) presents the detection strategy applied to a 2D data matrix (i.e., either a range-Doppler image, the range cross-range image, or the angle Doppler image).

The target detection problem can be expressed as

$$z_{\text{cut}} = \begin{cases} 1 & \text{if } z_{\text{cut}} > \mu\sigma_{cut}^2 \\ 0 & \text{if } z_{\text{cut}} < \mu\sigma_{cut}^2 \end{cases} \tag{1.45}$$

Radar detection algorithms are typically based on Neyman-Pearson (NP) criterion, wherein the probability of false alarm, P_{FA}, is set constant and the detector strategy is to maximize the probability of detection, P_D, for a given SNR in the presence of varying interference or noise levels. Such detectors are referred to as a constant false-alarm rate (CFAR) detector. In an indoor environment or automotive setting, varying clutter levels arise due to interfering objects in the field of view and multiple targets. The NP detector assumes that the interference is independent and identically distributed (IID) over all resolution cells and the noise variance of the interference distribution are estimated.

One of the simplest form of CFAR detection is the cell averaging CFAR algorithm, wherein the square value of the reference cells from the lagging cell and leading cells are added to estimate the noise variance for the cell under test;

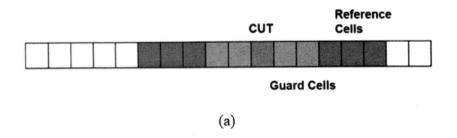

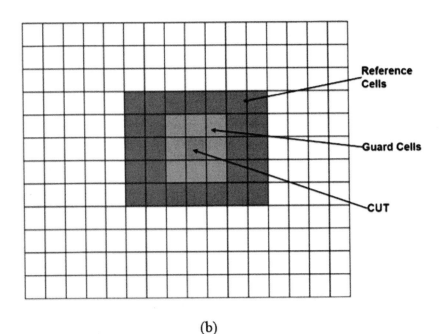

(b)

Figure 1.18 (a) 1D CFAR detector following 1D transformation, and (b) 2D CFAR detector following 2D transformation.

that is,

$$\sigma_{\text{cut}}^2(\text{CA}) = \sum_{r=1}^{N/2} |z_{\text{lagging}}(r)|^2 + \sum_{r=1}^{N/2} |z_{\text{leading}}(r)|^2 \tag{1.46}$$

The threshold multiplier μ for the CFAR detection is set by an acceptable probability of false alarm, which for cell-averaging CFAR (CA-CFAR) is provided as

$$\mu = N(P_{fa}^{-1/N} - 1) \tag{1.47}$$

where P_{fa} is the probability of false alarm and N is the window size used for noise power estimation.

In the case of CA-CFAR, there are guard cells that are adjacent cells to the CUT and are left out from computation of the noise variance for the threshold for CUT. The size of the reference cells and the guard cells are hyperparameters and are chosen to optimize target detection for a given false alarm rate. CA-CFAR performs well for a point target; however, in the case of range or Doppler spread targets the target scatterers are present in the reference cells and thus biasing the noise variance estimation to higher value and missed detection. Further, CA-CFAR elevates the noise threshold near a strong target, thus occluding or masking nearby weaker targets leading to a low probability of detection.

The alternative CFAR detectors are smallest of CA-CFAR (SOCA-CFAR) and greatest of CA-CFAR (GOCA-CFAR), which are given as

$$\sigma^2_{\text{cut}}(\text{SOCA}) = \min\{\sum_{r=1}^{N/2} |z_{\text{lagging}}(r)|^2, \sum_{r=1}^{N/2} |z_{\text{leading}}(r)|^2\}$$

$$\sigma^2_{\text{cut}}(\text{GOCA}) = \max\{\sum_{r=1}^{N/2} |z_{\text{lagging}}(r)|^2, \sum_{r=1}^{N/2} |z_{\text{leading}}(r)|^2\} \tag{1.48}$$

SOCA-CFAR and GOCA-CFAR are computationally cheaper than CA-CFAR and help in handling scenarios where the clutter and noise levels change abruptly. However, like their CA-CFAR counterpart, they suffer from the target masking and biased estimate issues in the case of extended targets.

Alternately for doubly spread targets, order-statistics CFAR (OS-CFAR) is used to avoid issues of biased estimates, since the ordered statistic is robust to any outliers. If the target's spread lies in the reference cells the noise estimate is not affected. In case of OS-CFAR, instead of the mean power in the reference cells, the kth ordered data is selected as the estimated noise variance, σ^2. A detailed description of OS-CFAR can be found in [15,16]. In the case of heterogeneous clutter where the noise parameters vary from one resolution cell to another, an adaptive clutter map is adaptively computed from frame to frame for noise statistics.

1.6.2 Clustering

Contrary to a point target, in the case of doubly extended targets, the output of the detection algorithm is not a single detection in the RDI for a target but is spread across range and Doppler. Thus, a clustering algorithm is required to group the detections from a single target, based on its size, as a single cluster. This helps in reducing the computational complexity for the target tracking algorithm, which after clustering tracks a single target parameter instead of tracking, nonclustered group of target parameters.

One of the simplest and effective clustering algorithms is based on the Euclidian distance metric. In the Euclidean distance-based clustering algorithm, the strongest detection is picked and the Euclidean distance between itself and the neighboring detections are computed. If they fall within the prior known object size they are clustered together as reflections from a single target and assigned a label. Subsequently, the next strongest detection is picked and the cluster is formed again based on the Euclidean distance lying within the objects geometric size and assigned another label. This is continued until all the detected targets are clustered and assigned a label. The drawback of this approach is that in a dense target reflection environment, cluster boundaries can be highly skewed by the target reflection strengths, which can result in incorrect assignments of detections to clusters. Further, this clustering algorithm requires the target object size as a a priori information.

A much more robust and effective unsupervised clustering algorithm is called the density-based spatial clustering of applications with noise (DBSCAN) algorithm [17]. Given a set of target detections from the same and multiple targets in the 2D space, DBSCAN groups detections that are closely packed together, while at the same time removing detections that lie alone in low-density regions as outliers. To do this, DBSCAN classifies each point as either a core point, edge point, or noise. Two input parameters are needed for the DBSCAN clustering algorithm—the neighborhood radius d, and the minimum number of neighbors, M. A point is defined as core point if it has at least $M - 1$ neighbors (i.e., points within the distance d). An edge point has less than $M - 1$ neighbors, but at least one of its neighbors is a core point. All points that have less than $M - 1$ neighbors and no core point as a neighbor do not belong to any cluster and will be classified as noise [17]. Figure 1.19 outlines the process of clustering target

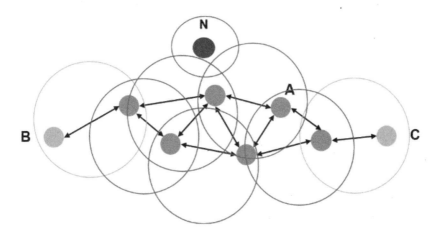

Figure 1.19 Illustration of DBSCAN algorithm.

Table 1.2
Confusion Matrix of True and Predicted Positive
and Negative Labels

		Predicted Class	
		Negative	**Positive**
Actual Class	**Negative**	True negative	False positive
	Positive	False negative	True positive

detections using the DBSCAN algorithm using the neighborhood radius d, and the minimum number of neighbors, M. OPTICS is an adaptive version of the DBSCAN algorithm where the parameters are adaptively computed.

Table 1.2 depicts an example of a confusion matrix indicating the actual and predicted classes for positive detection and negative detections. Accuracy, in this case, is computed as $\frac{\text{true positive}+\text{true negative}}{\text{total samples}}$; however, such a metric doesn't capture class imbalance issues and thus doesn't convey the complete performance picture.

One of the means of evaluating the detection and clustering algorithm is to use radar receiver operating characteristics (ROC) where for a given probability of false alarm (pFA) the probability of detection (pD) for a given SNR is computed and plotted. Figure 1.20 presents the ROC curves for two different detector-clustering algorithm combination and the objective of the algorithms is to maximize the area under the curve (AUC).

Figure 1.21 presents an exemplary setup, where a human target walking at 5 m and at an angle of $-20°$ is sensed in a small room with chairs, tables, furniture, and wall reflections, and Figure 1.22 presents the processing pipeline for that scenario. Range Doppler processing is first performed on the raw ADC data from a frame. Figure 1.22(a) presents the RDI of the scenario, where it can be observed that the actual human target is overshadowed by strong radar returns from static objects like chairs and the table present in the room. A 2D MTI filter is applied on RDIs from consecutive frames to remove reflections from static targets. The resulting RDI is presented in Figure 1.22(b). The filter is designed to be able to retain the micro-Doppler signatures of a human target and remove only zero Doppler targets, thus strong reflections from furniture in the room are removed from the RDI and the response is concentrated on the actual moving human target at range 5m. After range-Doppler processing, a range-angle image (RDI) is generated by range-angle processing, using techniques such as digital beam-forming, and combined to a single RAI through maximal ratio combining as shown in Figure 1.22(c). The 2D OS-CFAR is carried out on the RAI to mask out noise and obtain valid target detections as shown in Figure 1.22(d). Since

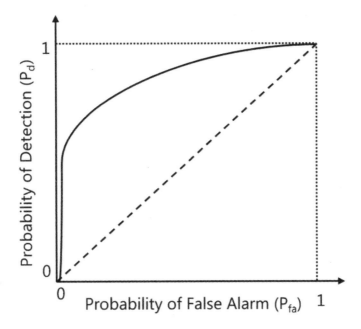

Figure 1.20   ROC performance of the detector and clustering algorithm.

Figure 1.21   An exemplary scenario, where a human is walking in a small room with furnitures and at 5 m and −20° from the sensor position.

human reflections at the high resolution radar's receiver exhibits high target spread along range and angle, and there are also outliers due to artifacts and noise on the RAI, DBSCAN helps to deal with such issues and is used to cluster the point cloud into target clusters. The effect of DBSCAN is presented in Figure 1.22(e), where the target is clearly observed at 5m and −20 degrees. These target clusters

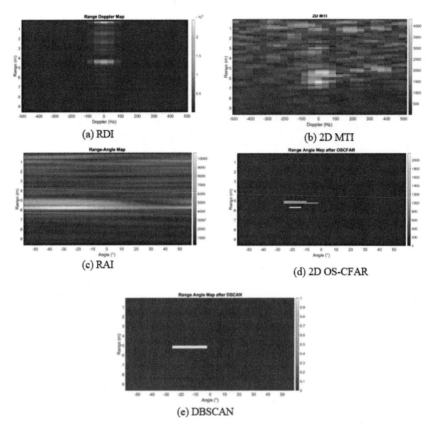

Figure 1.22   An exemplary processing pipeline, where an actual human target is at range 5 m and angle −20° to the sensor. (a) RDI, (b) RDI after 2D MTI filtering, (c) RAI by range-angle transformation followed by maximal ratio combining, (d) RAI after 2D OS-CFAR detector, and (e) RAI after DBSCAN clustering depicting that the target has been localized.

are then used to estimate the range, angle, and Doppler of the respective targets using the weighted mean of the cluster or centroid estimation.

1.7 Parameter Estimation and Cramer-Rao Bound

Once the target clusters are formed, the target parameters are obtained by taking a weighted sum of the range, Doppler, and/or angle of arrival values to estimate the effective target range, Doppler, and angle. Alternately, the target parameter can be obtained as the centroid of the cluster or the defined as the (min, max) values of the cluster. The performance of target parameter estimation is measured by resolution and accuracy. While resolution in either range, Doppler, or angle is defined by how close two point targets can be brought so that the radar is still able to distinguish them, accuracy measures the maximum variances of an unbiased estimator. Range resolution can be intuitively visualized as the 3-dB beamwidth extended along range by a point target at a certain distance at the radar receiver processing. The Doppler resolution and angle resolutions can be similarly estimated. The theoretical range, Doppler, and angle resolution have been defined earlier, and now the accuracy performance is presented in this section.

In general, the Cramér-Rao lower bounds on the variances of unbiased estimates of target delay τ_a and Doppler f_a can be expressed as

$$\operatorname{var} \theta \geq -\left[E\{L''(y(t); \tau_a, f_a)\}\right]^{-1}$$

$$= \left[\int y_{s,r}^{*\prime}(t; \tau_a, f_a)\, g'(t; \tau_a, f_a)\, dt\right]^{-1} \tag{1.49}$$

Here θ denotes τ_a or f_a, primes are derivatives with respect to θ. It needs to be pointed out that the denominator quantities on the right side in (1.49) are elements of the Fisher information matrix for the estimation problem. However, the right sides do represent the variance lower bound for any one of the parameters when the other parameter is known or considered fixed at a particular value.

In the case of matched filtering, (1.49) reduces to

$$\operatorname{var} \theta \geq \left[\int \left|y_{s,r}'(t; \tau_a, f_a)\right|^2 dt\right]^{-1}$$

$$= \left[\int \left|Y_{s,r}'(f; \tau_a, f_a)\right|^2 dt\right]^{-1} \tag{1.50}$$

ignoring all front end constants. In this case, the Cramér-Rao bound of range, on considering τ_a, defines the lower limit on accuracy with which the range to a target with known velocity can be measured

$$\sigma_r^2 \geq \frac{k}{SNR B_{rms}^2} \tag{1.51}$$

where k is constant, and $B_{rms} = \sqrt{\dfrac{\int_{-\infty}^{\infty} (2\pi f)^2 |S(f)|^2 df}{\int_{-\infty}^{\infty} |S(f)|^2 df}}$ denotes the root mean square bandwidth of the transmitted waveform. In the case of LFM, B_{rms} is equivalent to chirp bandwidth up to some constant due to its flat frequency response. This indicates that the range accuracy is a function of range resolution and the also decreases proportionally with the SNR.

Similarly, the Cramér-Rao bound of Doppler where the radar range is kept contant is provided as

$$\sigma_v^2 \geq \frac{k}{SNR \tau_{rms}^2} \tag{1.52}$$

up to some contant k, and τ_{rms} defined as $\sqrt{\frac{\int_{-\infty}^{\infty} t^2 |s(t)|^2 dt}{\int_{-\infty}^{\infty} |s(t)|^2 dt}}$ and is proportional to the chirp time in the case of a LFM.

1.8 Tracking

Once the target detections are available at each frame the next step is to associate them to existing tracks or create new tracks and delete tracks as necessary. The detected targets and their estimated parameters are prone to errors due to measurement noise, missed detections due to occlusions, false alarms due to ghost targets and interference sources. The task of tracking is to maintain and update the state and identity of these targets over time, reducing such false positive and true negatives. There are two important aspects of tracking. The first is the track management part, which handles track initiation, track maintenance, and measurement to track association. The other part is the track filtering step, which is achieved through either an alpha-beta filter, Kalman filter, or particle filter.

1.8.1 Track Management
1.8.1.1 Track Initiation
Track initiation is the process of creating a new radar track from an unassociated radar measurement. During track initialization, when a new unassociated radar measurement is encountered, the measurement is assigned to a new track, but the status of the track is kept tentative. The status of these tracks are changed to confirmed only after N out of M ($N < M$) subsequent measurements have a positive detection/measurement of the target. This reduces false positives or spurious targets from ghost targets and interferences to enter into the tracks.

1.8.1.2 Track Maintenance
Track maintenance is the process in which a decision is made if a track has to be terminated. If a track was not associated with a measurement during the association phase for consecutive N frames, there is a high likelihood that the target no longer exists in the radar's field of view and thus such tracks are terminated. Alternately, due to missed detection there are chances that the radar didn't detect that target for some measurements but would appear at a future frame time. In this case, the tracks are retained and only the prediction is updated.

1.8.1.3 Measurement to Track Association
The task of the measurement to track association is to assign the current measurement to an existing track. This step is also referred to as gating in the literature. Assuming at time step $n + 1$ there are N tracks; that is,

$$X = \{(\mu_1, \phi_1), (\mu_2, \phi_2), \cdots, (\mu_N, \phi_N)\} \tag{1.53}$$

where $\{\mu_i, \phi_i\}_{i=1}^N$ are the state mean and state error covariance matrices of the tracks. And the M measurements coming from target detections at time step $n+1$ are

$$Z = \{z_1, z_2, \cdots, z_M\} \tag{1.54}$$

A measurement to track association function $\Phi : Z \to X$ maps the uncertain measurements to certain tracks. The measurement to track association is a difficult problem due to multiple targets and the associated false positives and true negatives of the target detections. One of the simplest measurement to track the association algorithm is the suboptimal nearest neighbor (SNN), whereby each measurement is assigned to a track closest to it in a Euclidean or Mahanalobis distance sense. The assignment starts with the strongest detected target and thus is refered to as SNN.

The alternate approach includes the probabilistic data association filter (PDAF) or joint probabilistic data association filter (JPDAF).

1.8.2 Track Filtering

One of the simplest track smoothing filters are the alpha-beta filters. Alpha-beta filters do not require a detailed system model and rely on the moving averaging principle. The alpha-beta filter predicts position and velocity assuming zero acceleration of a moving target. It iterates the prediction as well as updates and smooths processes. The prediction process is expressed as follows:

$$\begin{cases} p_{n+1|n}^x = p_{n|n}^x + v_{n|n}^x \delta T \\ p_{n+1|n}^y = p_{n|n}^y + v_{n|n}^y \delta T \\ v_{n+1|n}^x = v_{n|n}^x \\ v_{n+1|n}^y = v_{n|n}^y \end{cases} \tag{1.55}$$

where δT is the measurement update interval, $(p_{n+1|n}^x, p_{n+1|n}^y)$ and $(v_{n+1|n}^x, v_{n+1|n}^y)$ are the predicted target position and velocity for time $n+1$ using data at time n in x- and y-axes, respectively, and $(p_{n|n}^x, p_{n|n}^y)$ and $(v_{n|n}^x, v_{n|n}^y)$ are the smoothed target position and velocity at time n in x- and y-axes, respectively.

The update and smoothing process is defined as

$$\begin{cases} p_{n+1|n+1}^x = p_{n+1|n}^x + \alpha(\hat{p}_{n+1}^x - p_{n+1|n}^x) \\ p_{n+1|n+1}^y = p_{n+1|n}^y + \alpha(\hat{p}_{n+1}^y - p_{n+1|n}^y) \\ v_{n+1|n+1}^x = v_{n+1|n}^x + \beta(\hat{v}_{n+1}^x - v_{n+1|n}^x) \\ v_{n+1|n+1}^y = v_{n+1|n}^y + \beta(\hat{v}_{n+1}^y - v_{n+1|n}^y) \end{cases} \tag{1.56}$$

where $(\hat{p}_{n+1}^x, \hat{p}_{n+1}^y)$ and $(\hat{v}_{n+1}^x, \hat{v}_{n+1}^y)$ are the measured position and velocity of the target at $(n+1)\delta T$ refresh time, respectively, in the x- and y-axis. The parameters (α, β) are chosen to be fixed independent of the data. As can be seen,

$\hat{p}_{n+1}^{x} - p_{n+1|n}^{x}$ is the amount of distance along the x-axis, which hasn't been accounted by the prediction $p_{n+1|n}^{x}$ by the new measurement $\hat{p}_{n+1}^{x}$.

Alternately, the Kalman filter is an algorithm that utilizes a series of noisy measurement observed over time to estimate the internal state of a linear dynamic system. The estimated state variables are more accurate than the alpha-beta filter since the algorithm estimates a joint probability distribution over the variables for each time frame. The algorithm has two steps. In the prediction step, the filter predictes the next internal state based on the process model accounting for the model uncertainties. In the measurement step, the measurement is combined with the predicted state in an adpative weighted average referred to as Kalman gain to update the internal state estimate and its uncertainty. The Kalman filter operates under the Bayesian principle, so when the measurement data is observed with less uncertainty compared to the predicted data the Kalman gain is high, which means it relies more on the measured data to update state. On the other hand, when the predicted data has less uncertainty compared to that of measurement data the Kalman gain is low, meaning it weighs the predicted data more for its state update. Kalman filters are a generic algorithm that find use in several signal processing scenarios to control theoretic applications and are not limited to radar signal processing. The process model used by the Kalman filter is similar to that of a hidden Markov model except that the state space is rather continuous and assumes the state and observed variables to follow normal distribution.

One of the most standard used process models or state transition functions is a linear constant velocity (CV) motion model. Let the state variable be $\psi = \begin{bmatrix} p^x & p^y & v^x & v^y \end{bmatrix}^T$, where p^x, p^y denote the position coordinates in x and y coordinates, whereas v^x, v^y denote the velocity in the x- and y-axes, respectively. The CV process model can be expressed as

$$\begin{cases} p_{n+1|n}^{x} = p_{n|n}^{x} + v_{n|n}^{x}\delta t + n_{px}^{p} \\ p_{n+1|n}^{y} = p_{n|n}^{y} + v_{n|n}^{y}\delta t + n_{py}^{p} \\ v_{n+1|n}^{x} = v_{n|n}^{x} + n_{vx}^{p} \\ v_{n+1|n}^{y} = v_{n|n}^{y} + n_{vy}^{p} \end{cases} \qquad (1.57)$$

where the state variables with suffix $n|n$ denotes the a posteriori probability and $n + 1|n$ denotes the a priori probability. δt represents the frame time (i.e., the radar refresh rate). n^p denotes the Gaussian process noise and captures the kinematic changes, which are not considered in the linear process update model. In the compact matrix-vector form, the process model can be expressed as

$$\begin{bmatrix} p_{n+1|n}^{x} \\ p_{n+1|n}^{y} \\ v_{n+1|n}^{x} \\ v_{n+1|n}^{y} \end{bmatrix} = \begin{bmatrix} 1 & 0 & \delta t & 0 \\ 0 & 1 & 0 & \delta t \\ 0 & 0 & 1 & 0 \\ 0 & 0 & 0 & 1 \end{bmatrix} \begin{bmatrix} p_{n|n}^{x} \\ p_{n|n}^{y} \\ v_{n|n}^{x} \\ v_{n|n}^{y} \end{bmatrix} + \begin{bmatrix} n_{px}^{p} \\ n_{py}^{p} \\ n_{vx}^{p} \\ n_{vy}^{p} \end{bmatrix} \qquad (1.58)$$

that is, $\psi_{n+1|n} = F\psi_{n|n} + n^p$. Now since the state variables are assumed to be Gaussian distributed, the process noise is also modeled as zero mean Gaussian noise, thus the mean and covariance of the state variables can be expressed as

$$\mu_{n+1|n} = E\{\psi_{n+1|n}\} = F\mu_{n|n}$$

$$\phi_{n+1|n} = E\{\phi_{n+1|n}\phi^T_{n+1|n}\} = F\phi_{n|n}F^T + Q \tag{1.59}$$

where $E\{.\}$ is the expectation operator and Q denotes the process noise covariance matrix. In the case of the constant velocity model, the process noise would include kinematic changes due to jerks and acceleration of the target. Considering the acceleration component, the noise process can be expressed as

$$n^p = \begin{bmatrix} \frac{1}{2}a_x\delta t^2 \\ \frac{1}{2}a_y\delta t^2 \\ a_x\delta t \\ a_y\delta t \end{bmatrix} = \begin{bmatrix} \frac{1}{2}\delta t^2 & 0 \\ 0 & \frac{1}{2}\delta t^2 \\ \delta t & 0 \\ 0 & \delta t \end{bmatrix} \begin{bmatrix} a_x \\ a_y \end{bmatrix} = Ga \tag{1.60}$$

Thus the process noise covariance matrix is defined as

$$Q = E\{n^p n^{pT}\} = G\begin{bmatrix} a_x^2 & 0 \\ 0 & a_y^2 \end{bmatrix} G^T \tag{1.61}$$

where a_x^2, a_y^2 represents the variance of acceleration noise in the x- and y-axis, and the covariance acceleration noise in the x- and y-axis are 0. Thus the process noise covariance Q is initialized based on the maximum target's acceleration expected in the system.

In the next step, the measurement model needs to account for the transformation of the internal state to the radar measurements (i.e., the radial distance ρ), the azimuth angle or bearing angle θ, and the radial velocity or range rate v.

$$\begin{cases} \rho = \sqrt{p^{x2} + p^{y2}} \\ v = \frac{p^x v^x + p^y v^y}{\sqrt{p^{x2} + p^{y2}}} \\ \theta = \tan^{-1}(\frac{p^y}{p^x}) \end{cases} \tag{1.62}$$

that is, $z = H(\psi) + n^m$, where n^m represents the measurement zero-mean Gaussian noise with covariance matrix $R = E\{n^m n^{mT}\}$. This indicates that unlike the process model, the measurement model, which maps the predicted state $\begin{bmatrix} p^x & p^y & v^x & v^y \end{bmatrix}^T$ to the measurement space $z = \begin{bmatrix} \rho & v & \theta \end{bmatrix}^T$, is a nonlinear transformation, which implies the simple Kalman filter cannot be applied in this case. The two common approaches to handle such a nonlinear transformation issue are the extended Kalman filter and the unscented Kalman filter. While in the extended Kalman filter the nonlinear equation is approximated through a first-order Taylor's expansion, unscented Kalman filter uses sigma points to sample from the a priori state distribution, which is then propagated through the nonlinear transformation, and then its mean and covariance are estimated, which forms the transformed state.

1.8.2.1 Extended Kalman Filter

In the extended Kalman filter [18], the nonlinear transformation is approximated by a first-order Taylor's expansion as

$$H(\mu) = H(\mu_0) + J_{H(\mu)}(\mu - \mu_0) \tag{1.63}$$

where $J_{H(\mu)}$ is the Jacobian matrix of H with regard to state variable ψ; that is,

$$J_{H(\mu)} = \begin{bmatrix} \frac{\partial \rho}{\partial p^x} & \frac{\partial \rho}{\partial p^y} & \frac{\partial \rho}{\partial v^x} & \frac{\partial \rho}{\partial v^y} \\ \frac{\partial v}{\partial p^x} & \frac{\partial v}{\partial p^y} & \frac{\partial v}{\partial v^x} & \frac{\partial v}{\partial v^y} \\ \frac{\partial \theta}{\partial p^x} & \frac{\partial \theta}{\partial p^y} & \frac{\partial \theta}{\partial v^x} & \frac{\partial \theta}{\partial v^y} \end{bmatrix} \tag{1.64}$$

which after the partial derivatives reduces to

$$J_{H(\mu)} = \begin{bmatrix} \frac{p^x}{\sqrt{p^{x2}+p^{y2}}} & \frac{p^y}{\sqrt{p^{x2}+p^{y2}}} & 0 & 0 \\ \frac{p^y(v^x p^y - v^y p^x)}{(p^{x2}+p^{y2})^{3/2}} & \frac{p^x(v^y p^x - v^x p^y)}{(p^{x2}+p^{y2})^{3/2}} & \frac{p^x}{\sqrt{p^{x2}+p^{y2}}} & \frac{p^y}{\sqrt{p^{x2}+p^{y2}}} \\ -\frac{p^y}{p^{x2}+p^{y2}} & \frac{p^x}{p^{x2}+p^{y2}} & 0 & 0 \end{bmatrix} \tag{1.65}$$

Then, given the measurement $\hat{z}_{n+1}$ the innovation and its covariance are computed as

$$\begin{cases} i_{n+1} = \hat{z}_{n+1} - J_{H(\mu)} x_{n+1|n} \\ \Omega_{n+1} = J_{H(\mu)} P_{n+1|n} J_{H(\mu)}^T + R \end{cases} \tag{1.66}$$

The innovation and its covariance represent the additional information made available on receipt of the measurement and not captured by the process update step. The Kalman gain is computed as

$$K_{n+1} = \phi_{n+1|n} J_{H(\mu)}^T \Omega_{n+1}^{-1} \tag{1.67}$$

Then the posterior state variables' mean and covariance are updated as

$$\mu_{n+1|n+1} = \mu_{n+1|n} + K_{n+1} i_{n+1}$$
$$\phi_{n+1|n+1} = (I - K_{n+1} J_{H(\mu)}) \phi_{n+1|n} \tag{1.68}$$

1.8.2.2 Unscented Kalman Filter

Due to the first-order approximation, an extended Kalman filter (EKF) can introduce large errors in the estimated posterior mean and covariance, which potentially can lead to suboptimal performance and in several cases cause divergence of the filter. An unscented Kalman filter (UKF) [19] attempts to solve this issue using deterministic sampling of so called 'sigma points' from the prior distribution and thus when propagated through the nonlinear transform leads to accurate mean and covariance estimation of the posterior distribution up to second order.

$$\begin{cases} x_0 = \mu_{n+1|n} \\ x_i = \mu_{n+1|n} + col_i(\sqrt{(\lambda + N)\phi_{n+1|n}}) & ; i = 1, \cdots, N \\ x_i = \mu_{n+1|n} - col_i(\sqrt{(\lambda + N)\phi_{n+1|n}}) & ; i = L+1, \cdots, 2N \\ w_0 = \frac{\lambda}{N+\lambda} \\ w_i = \frac{1}{2(N+\lambda)} & ; i = 1, \cdots, 2N \end{cases} \quad (1.69)$$

where $col_i(.)$ operation picks the ith column from the matrix and N is the dimension of the state variable. λ is the sigma point spreading parameter and is usually kept as 3. These sigma points sampled from the prior distribution $\mathcal{N}(\mu_{n+1|n}, \phi_{n+1|n})$ are then passed through the nonlinear transformation to produce the transformed sigma points, $y_i = H(x_i); i = 0, 1, \cdots, 2N$. Then, the mean and covariance of the transformed distribution in measurement space is estimated by the weighted average of their transformed sigma points as

$$z_{n+1|n} = \sum_{i=0}^{2N} w_i y_i$$

$$\Omega_{n+1} = \sum_{i=0}^{2N} w_i (y_i - z_{n+1|n})(y_i - z_{n+1|n})^T + R \quad (1.70)$$

Note that the parameter Ω_{n+1} is the innovation covariance matrix.

The cross-covariance between the measurement domain z and the state domain ψ is computed as

$$\Upsilon_{n+1} = \sum_{i=0}^{2N} w_i (y_i - z_{n+1|n})(x_i - \mu_{n+1|n})^T \quad (1.71)$$

Then, the Kalman gain K_{n+1} and the mean $\mu_{n+1|n+1}$ and covariance $\phi_{n+1|n+1}$ of the posterior distribution are computed as

$$\begin{cases} K_{n+1} = \Upsilon_{n+1}\Omega_{n+1}^{-1} \\ \mu_{n+1|n+1} = \mu_{n+1|n} + K_{n+1}(\hat{z}_{n+1} - z_{n+1|n}) \\ \phi_{n+1|n+1} = \phi_{n+1|n} - K_{n+1}\hat{\phi}_{n+1|n}K_{n+1}^T \end{cases} \quad (1.72)$$

where $\hat{z}_{n+1}$ is the measurement coming from the target detection at frame time $n + 1$.

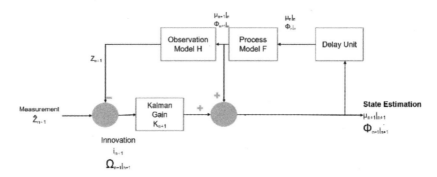

Figure 1.23 Pictorial Representation of the overall Kalman Filter for tracking radar targets.

The above unscented transformation requires several orders of magnitude less sampling points compared to that required in Monte Carlo simulations or particle filters, mainly due to the Gaussian distribution assumption. Figure 1.23 presents the Kalman filter loop depicting the prior state estimates, innovation estimate and the posterior state estimates.

Figure 1.24 presents a scenario where two people walk into the room, one at about 5m and another at 3m, and then are subsequently walking radially along their radial angles with respect to the radar sensor, Figure 1.24(a) presents the detections over frames, which include misdetections and also detections from ghost targets. Figure 1.24(b) presents the output of the tracker where the two people are identified along their range, Doppler, and angle parameters. The tracks depict no swap between the tracked targets in this scenario and no false tracks being generated and no lost tracks compared to the ground truth. Figure 1.25 presents a scenario where one person start walking from 6m towards the sensor and another person enters close to the sensor and walks away from the sensor. They both perform criss-cross walks along the range in a small meeting room with tables, chairs, and a wall at 6m. Figure 1.24(a) presents the detections over frames, which include misdetections and also detections from ghost targets. Figure 1.24(b) presents the output of the tracker where the two people are identified along their range, Doppler, and angle parameters. Around frames 190 to 240, one target is lost as he moves out of the field of view and reenters. A track discontinuation and reappearance of the track is seen at the output of the tracker for the scenario.

One of the major challenges of Kalman filters is setting the correct process noise and measurement noise parameters. If these parameters are not set correctly the Kalman filter can diverge. The measurement noise variance along range, Doppler, and angle are typically specified by the radar specification and thus is still relatively easier to set. The process noise on the other hand is application- and object-dependent and requires to be set adaptively. One of the means to determine if the noise parameters of the Kalman filter are set appropriately is the

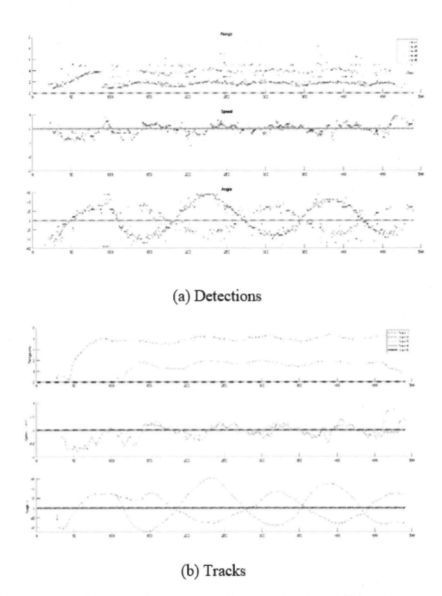

(a) Detections

(b) Tracks

Figure 1.24 (a) Target detections over range, Doppler, and angle, and (b) target tracks over range, Doppler, and angle for two people at a range of about 3 m and 5 m from the sensor walking radial movement.

normalized innovation squares (NISs) metric defined as

$$\text{NIS}_{n+1} = i_{n+1}\Omega_{n+1}^{-1}i_{n+1}^{H} \tag{1.73}$$

NIS_{n+1} is χ^2- distributed with m degree of freedom, where m is the number of measurement variables.

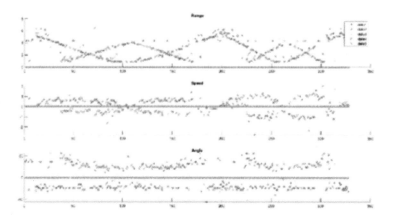

(a) Detections

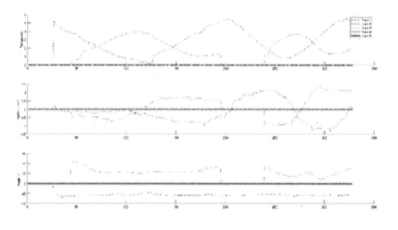

(b) Tracks

Figure 1.25 (a) Target detections over range, Doppler, and angle, and (b) target tracks over range, Doppler, and angle for two people walking criss-cross close and away from the sensor.

When the NIS is large and falls outside the 95% confidence region it is an indication that the process and measurement noise variance are set lower than the actual levels of the system. Figure 1.26(a) illustrates the scenario when the NIS is larger than the 95% confidence interval. Intuitively this means that the new measurement is always outside the gating function around the predicted state

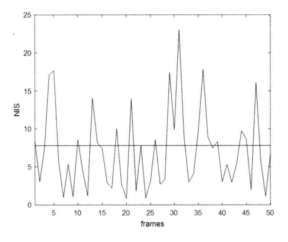

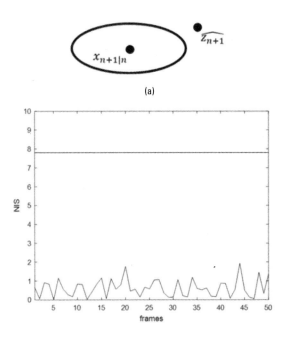

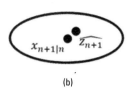

Figure 1.26 (a) NIS is higher than 95% confidence score, and (b) NIS is lower than 5% confidence score.

variable, implying that we underestimated the noise values. This is a bad situation since it implies that the prediction isn't accurate. When NIS is small and falls below the 5% confidence region, it is an indication that the process and measurement noise variance are set higher than the actual levels of the system. Figure 1.26(b) illustrates the scenario when the NIS is smaller than the 5% confidence interval. Intuitively this means that the new measurement is always very close to the predicted state variable compared to the gating function, implying that we overestimated the noise values, which indicates that the measurement doesn't bring in new information that isn't already available through the prediction step.

1.9 Applications of Short-Range Radar

Automotive radars with highly integrated mm-wave circuits implementation have become a common feature for automotive safety and control. Both self-driving autonomous cars and human-driven cars are increasingly using radars to improve passengers' comfort and safety. For instance, forward-looking automatic cruise control radars [20] are long-range radars that enable the vehicle to automatically adjust the vehicle speed according to the preceding vehicle. Blind spot detection systems and lane change assist radars [21] are medium-range radars that facilitate safer driving by warning a potential collision to cars in blind spots or if a lane change requires a dangerous maneuver [22–24]. Parking assist and rear cross traffic alert radars are short-range radars that help drivers during parking or navigating through parking lots. Figure 1.27(a) shows typical automotive radars fitted in modern cars for such safety and control applications. Apart from advanced driving assistance system (ADAS) safety, new realms of applications for in-cabin safety and comfort are also gaining attention for enabling a holistic autonomous driving experience. The applications for ultrashort-range radar are proliferating, be it radar-enabled gesture-controlled dashboard to facilitate the driver's attention on road or reminding a parent of their forgotten child in a car. Figure 1.27(b) depcits the emerging applications of dooring to alert the passenger and avoid opening of the door in case of a passing bike or motorcycle, gesture sensing in dashboard for controlling infotainment systems, occupancy sensing to sense if there are people left behind in the car, and driver monitoring radar systems.

Wide adoption of short-range radar sensors in industrial and consumer applications demand reliable system performance at low power consumption, low cost, and with a small footprint. Radar adds several benefits to existing applications as well as provides new features allowing for completely new applications. In 3D localization in home and digital applications, radar provides range and angle in azimuth and elevation. Radar can sense velocity information, which can be utilized for position mapping and tracking. Further, radars can detect human cardiopulmonary motion, which provides a promising approach to overcome the problems of false trigger and dead spots in conventional sensors for occupancy

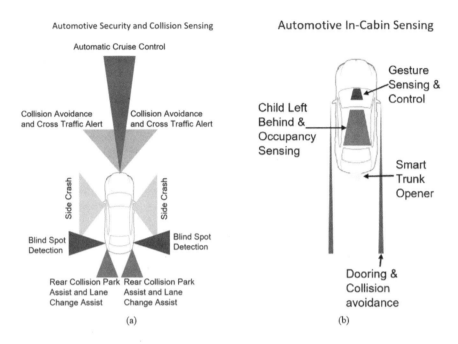

Figure 1.27 (a) Automotive radars for safety and control, and (b) automotive radars for in-cabin sensing and short-range sensing.

sensing applications. Radar technology returns a unique signature from any object or material. This feature can be utilized in systems to recognize different types of liquids, such as water vs. milk, or materials like silk vs. cotton. Thus radar system can enable low-cost solutions for industrial testing and automation. The robustness of radar to harsh environmental conditions, such as poor lighting, fog, or unclean sensors, promotes radars as a powerful and reliable sensor. Additionally, radar sensors can be aesthetically concealed without affecting performance, making them good candidates for consumer applications.

Gesture sensing is an effective and intuitive human-computer interface compared to the more traditional mouse and keyboard interface. It is an asset to have gesture sensing in a car dashboard to control the infotainment or navigation systems while allowing the driver to concentrate fully on driving without having to worry about the exact location of the buttons [25–28]. Further, remote controlling electrical appliances such as televisions and personal digital assistants through gestures offer flexibility and ease-of-use to its users. In addition, in augmented reality and virtual reality systems gesture control offers a valuable user feature. Figure 1.28 depicts how gesture sensing and control can be utilized in automotive and consumer applications.

In the United States, residential and commercial users account for 42% of total energy consumption. While in residential areas, about 30% of electricity is

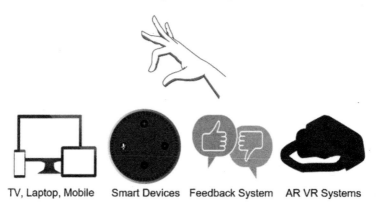

Gesture Recognition,Sensing ,Control

TV, Laptop, Mobile   Smart Devices   Feedback System   AR VR Systems

Figure 1.28   Gesture sensing, recognition, and control enable alternate human-computer interfaces, controlling devices remotely, and also faciliates augmented reality–virtual reality systems.

used for lighting and heating, ventilation, and air conditioning (HVAC) systems, in commercial buildings, they account for over 50% of electricity consumption, and in hotels they can account for up to 80% of the utility bill. Henceforth, large residential, commercial, and public spaces are inefficient when it comes to energy consumption. This demands the use of efficient occupancy sensors and people counting sensor systems that can facilitate intelligent and effective use of energy resources. Energy consumption can be significantly reduced in residential, commercial, or public spaces by monitoring occupancy or counting the number of people and accordingly regulating artificial light and HVAC systems [29,30]. Short-range radars can sense breathing and heartbeats from a distance, faciliating a ubiquitous occupancy sensor [31–35]. Figure 1.29(a) shows how radar can sense macro-Doppler while walking, runnning, exercising, and so on, micro-Doppler while working on a desktop, cooking, and so on, and vital-Doppler while sleeping, watching TV, and so on [36]. Further, Figure 1.29(b) depicts that an indoor radar sensor can offer a robust sensing modality to sense activities and elderly fall motions and be used to alert the authorities in case of emergencies [37–39].

Studies have suggested that a majority of drivers spend anywhere between 3.5 to 14 minutes in a typical search for a parking space [40]. These times quickly add up to cause significant productivity losses in cities. In addition to annoyance to individual drivers, cruising for parking impacts the overall efficiency of the transport system, introducing traffic snarls, slowdowns, and so on. Thus, as Figure 1.30 illustrates smart street lighting with mounted radars sensors can not only help in saving energy by switching off when there is no traffic near

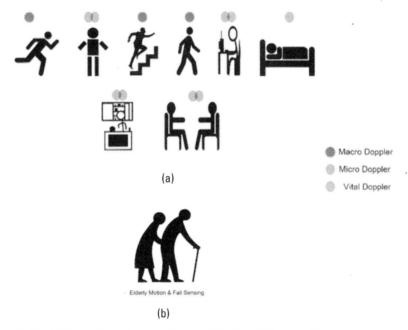

(a)

● Macro Doppler

◉ Micro Doppler

◉ Vital Doppler

Elderly Motion & Fall Sensing

(b)

Figure 1.29   (a) Human behavior modeling, and (b) elderly fall recognition.

**Street Lighting Control and Traffic Congestion
Control and Parking Count Alert**

Figure 1.30   Smart street lighting to sense cars and also signal a parking space.

its vicinity, but also help through centralized or distributed mechanism to alert drivers of empty or occupied parking near by.

Autonomous driving requires scene understanding, which means accurate classification of objects in and around the vehicle. Radars due to their all-weather feature are particularly attractive for advanced driver-assistance systems

| Car | Pedestrian | Bike | Motorcycle | Truck |

Figure 1.31 Object classification such as pedestrian, car, motorcycle, bike, and truck.

People Counting Applications

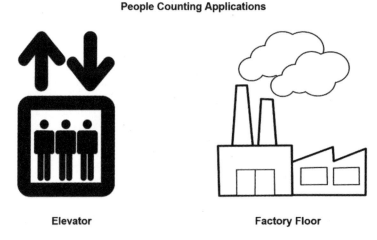

Elevator   Factory Floor

Figure 1.32 People/object localization in 3D space and people counting on elevator and industry floors.

(ADAS) and autonomous driving applications. Therefore, reliable classification of object labels such as pedestrian, car, motorbike, bike, truck, fence, or bush, especially by interesting features. Figure 1.31 illustrates some of these object classifications types. Further, object detection and classification through an automotive occupancy grid [41,42] helps in identifying open spaces, parking spaces, and so on.

Further, several applications require precise localization of objects in 3D space such as outdoor intruder notification or waking other power-hungry sensors by detecting users in the vicinity [43]. In other applications, an overall count of people may be required, such as counting people in an elevator, crowded space, or factory floor for monitoring, security, and accounting purposes without explicit localization of these targets in space, as illustrated in Figure 1.32.

Figure 1.33 presents a representation of a smart home and city, where several applications are automated and enabled through short-range radar technology. Unlike a vision-based system, a radar sensor does not need an opening or a lens. It can sense through many materials, thus saving cost and preserving its aesthetic

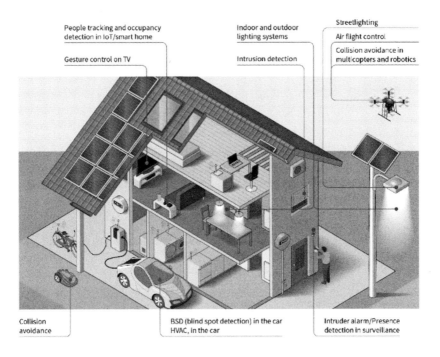

People tracking and occupancy detection in IoT/smart home

Gesture control on TV

Indoor and outdoor lighting systems

Intrusion detection

Streetlighting

Air flight control

Collision avoidance in multicopters and robotics

Collision avoidance

BSD (blind spot detection) in the car
HVAC, in the car

Intruder alarm/Presence detection in surveillance

Figure 1.33   Smart home and city enabled through smart technologies such as short-range radars.

value, while enabling new product designs and guaranteeing privacy. Several other novel personal medical applications enabled through short-range radars include noninvasive glucose monitoring [44] and biometric authentications [45].

1.10 Problems

1. Fourier equivalence dictates that a signal limited in time domain T is elongated in bandwidth B and vice versa. However, the time and bandwidth can be set independent of each other for defining chirp waveform. How can you justify the same?
2. What are the pros and cons of short-range FMCW radar, short-range UWB, radar and short-range OFDM radar?
3. What is the need for staggered PRT in upchirp downchirp waveform sequence? Why are longer chirp times (compared to upchirp waveforms) required for sequence of upchirp downchirp waveform?
4. What is the significance of the ambiguity function for waveform design? From the first principle, derive the (self) ambiguity function to include angle offsets in the standard Woodward ambiguity function.

5. When is the assumption of matched filtering, where the filter is matched to transmit waveform, not valid?

6. State an advantage and disadvantage of TDM, FDM, and slow-time CDM MIMO radar modes.

7. Explain the processing pipeline of 2D angle-Doppler image-based detection. Think of a use case, where 2D angle-Doppler image processing is preferred over other 2D slices.

8. What is the advantage of maximal ratio combining-based coherent integration over mean combining-based coherent integration? Explain coherent and noncoherent integration with respect to 1D radar data and 2D radar images.

9. When is slow-time FFT equivalent to the matched filter in that domain? Is slow-time FFT equivalent to the matched filter in the case of human targets? Why?

10. What is the difference between resolution and accuracy? State the resolution and accuracy limits for range, Doppler, and angle estimates with respect to other radar parameters.

11. What are the adaptation required in a processing pipeline for sensing extended targets compared to point target processing?

12. What are the advantages of an OS-CFAR detector over a CA-CFAR detector for extended targets? What is the need for guard cells in CA-CFAR? Why are they not absolutely necessary for OS-CFAR?

13. What are the advantages and disadvantages of DBSCAN and Euclidean clustering?

14. What does Cramer-Rao bound for estimating a radar parameter indicate? Derive the Cramer-Rao bound for angle estimation.

15. How is unscented transformation different than Monte-Carlo sampling, as used in a particle filter for example?

16. Derive the process model and measurement model of an unscened Kalman filter for a constant acceleration model.

17. How can a system engineer tweak process noise covariance matrix and measurement noise covariance matrix if they are unknown?

References

[1] Charvat, G. L., *Small and Short-Range Radar Systems*, Boca Raton, FL: CRC Press, 2014.

[2] Hillger, P., R. Jain, J. Grzyb, et al., "A 128-Pixel 0.56 THz Sensing Array for Real-Time Near-Field Imaging in 0.13 m SiGeBiCMOS," *2018 IEEE International Solid - State Circuits Conference -(ISSCC)*, 2018, pp. 418–420.

[3] http://www.tsmc.com/english/dedicatedFoundry/technology/logic.htm.

[4] Fujibayashi, T., Y. Takeda, W. Wang, et al., "A 76- to 81-GHz Multi-Channel Radar Transceiver," *IEEE Journal of Solid-State Circuits*, Vol. 52, No. 9, 2017, pp. 2226–2241.

[5]   Nasr, I., R. Jungmaier, A. Baheti, et al., "A Highly Integrated 60 GHz 6-Channel Transceiver with Antenna in Package for Smart Sensing and Short-Range Communications," *IEEE Journal of Solid-State Circuits*, Vol. 51, No. 9, 2016, pp. 2066–2076.

[6]   Bock, J., K. Aufinger, S. Boguth, et al., "SiGe HBT and BiCMOS Process Integration Optimization within the DOTSEVEN Project," *2015 IEEE Bipolar/BiCMOS Circuits and Technology Meeting-BCTM*, 2015, pp. 121–124.

[7]   Shopov, S., M. G. Girma, J. Hasch, N. Cahoon, and S. P. Voinigescu. "Ultralow-Power Radar Sensors for Ambient Sensing in the V-Band," *IEEE Transactions on Microwave Theory and Techniques*, Vol. 65, No. 12, 2017, pp. 5401–5410.

[8]   Hagelauer, A., M. Wojnowski, K. Pressel, R. Weigel, and D. Kissinger, "Integrated Systems-in-Package: Heterogeneous Integration of Millimeter-Wave Active Circuits and Passivesin Fan-Out Wafer-Level Packaging Technologies," *IEEE Microwave Magazine*, Vol. 19, No. 1, 2018, pp.48–56.

[9]   Ng, H. J., W. Ahmad, M. Kucharski, J. H. Lu, and D. Kissinger. "Highly-Miniaturized 2-Channel mm-Wave Radar Sensor with On-Chip Folded Dipole Antennas," *2017 IEEE Radio Frequency Integrated Circuits Symposium (RFIC)*, 2017, pp. 368–371.

[10] Blunt, S. D., and Mokole, E. L., "Overview of Radar Waveform Diversity," *IEEE Aerospace and Electronic Systems Magazine*, Vol. 31, No. 11, 2016, pp. 2–42.

[11] Santra, A., R. Srinivasan, K. Jadia, and G. Alleon, "Ambiguity Functions, Processing Gains, and Cramer-Rao Bounds for Matched Illumination Radar Signals," *IEEE Transactions on Aerospace and Electronic Systems*, Vol. 51, No. 3, 2015, pp. 2225–2235.

[12] Santra, A., A. R. Ganis, J. Mietzner, and V. Ziegler, "Ambiguity Function and Imaging Performance of Coded FMCW Waveforms with Fast 4D Receiver Processing in MIMO Radar," *Digital Signal Processing*, Vol. 97, 2020, p. 102618.

[13] Potter, L. C., D. M. Chiang, R. Carriere, and M. J. Gerry, "A GTD-Based Parametric Model forRadar Scattering," *IEEE Transactions on Antennas and Propagation*, Vol. 43, No. 10, 1995, pp. 1058–1067.

[14] Jackson, J. A., B. D. Rigling, and R. L. Moses, Canonical Scattering Feature Models for 3D and Bistatic SAR," *IEEE Transactions on Aerospace and Electronic Systems*, Vol. 46, No. 2, 2010, pp. 525–541.

[15] Blake, S., "OS-CFAR Theory for Multiple Targets and Nonuniform Clutter," *IEEE Transactions on Aerospace and Electronic Systems*, Vol. 24, No. 6, 1988, pp. 785–790.

[16] Gandhi, P. P., and S. A. Kassam, "Analysis of CFAR Processors in Nonhomogeneous Background," *IEEE Transactions on Aerospace and Electronic Systems*, Vol. 24, No. 4, 1988, pp. 427–445.

[17] Ester, M., H.-P. Kriegel, J. Sander, et al., eds., "A Density-Based Algorithm for Discovering Clusters in Large Spatial Databases with Noise," in *Proceedings of the Second International Conference on Knowledge Discovery and Data Mining (KDD-96)*.

[18] Anderson, B. D. O., and J. B. Moore, *Optimal Filtering*, Englewood Cliffs, NJ: Prentice Hall, 1979.

[19] Wan, E. A., and R. Van Der Merwe, "The Unscented Kalman Filter for Nonlinear Estimation," in *Proceedings of the IEEE 2000 Adaptive Systems for Signal Processing, Communications, and Control Symposium*, pp. 153–158.

[20] Wenger, J., "Automotive Radar: Status and Perspectives," in *Proc. IEEE Compound Semiconductor Integrated Circuit Symp.*, Palm Springs, CA, 2005, pp. 21–25.

[21] Hasch, J., E. Topak, R. Schnabel, T. Zwick, R. Weigel, and C. Waldschmidt, "Millimeter-Wave Technology for Automotive Radar Sensors in the 77 GHz Frequency Band," *IEEE Transactions on Microwave Theory and Techniques*, Vol. 60, No. 3, March 2012, pp. 845–860.

[22] Patole, S. M., M. Torlak, D. Wang, and M. Ali, "Automotive Radars: A Review of Signal Processing Techniques," *IEEE Signal Processing Magazine*, Vol. 34, No. 2, March 2017, pp. 22–35.

[23] Jones, W. D., "Keeping Cars from Crashing," *IEEE Spectrum*, Vol. 38, No. 9, 2001, pp. 40–45.

[24] Meinel, H. H., "Evolving Automotive Radar: From the Very Beginnings into the Future," *8th IEEE European Conference on Antennas and Propagation (EuCAP)*, April 2014, pp. 3107–3114.

[25] Lien, J., N. Gillian, M. E. Karagozler, et al., "Soli: Ubiquitous Gesture Sensing with Millimeter Wave Radar," *ACM Transactions on Graphics (TOG)* Vol. 35, No. 4, 2016, p. 142.

[26] Wang, S., J. Song, J. Lien, I. Poupyrev, and O. Hilliges, "Interacting with Soli: Exploring Fine-Grained Dynamic Gesture Recognition in the Radio-Frequency Spectrum," in *Proceedings of the 29th Annual Symposium on User Interface Software and Technology*, ACM, 2016, pp. 851–860.

[27] Smith, K. A., C. Csech, D. Murdoch, and G. Shaker, "Gesture Recognition Using mm-Wave Sensor for Human-Car Interface," *IEEE Sensors Letters*, Vol. 2, No. 2, 2018, pp. 1–4.

[28] Hazra, S., and A. Santra, "Robust Gesture Recognition Using Millimetric-Wave Radar System," *IEEE Sensors Letters*, Vol. 2, No. 4, 2018, pp. 1–4.

[29] Yavari, E., C. Song, V. Lubecke, and O. Boric-Lubecke. "Is There Anybody in There?: Intelligent Radar Occupancy Sensors," *IEEE Microwave Magazine*, Vol. 15, No. 2, 2014, pp. 57–64.

[30] Santra, A., R. V. Ulaganathan, and T. Finke, "Short-Range Millimetric-Wave Radar System for Occupancy Sensing Application," *IEEE Sensors Letters*, Vol. 2, No. 3, September 2018.

[31] Li, C., V. M. Lubecke, O. Boric-Lubecke, and J. Lin, "A Review on Recent Advances in Doppler Radar Sensors for Noncontact Healthcare Monitoring," *IEEE Transactions on Microwave Theory and Techniques*, Vol. 61, No. 5, 2013, pp. 2046–2060.

[32] Gu, C., "Short-Range Non contact Sensors for Health care and Other Emerging Applications: A Review," *Sensors*, Vol. 16, No. 8, 2016, p. 1169.

[33] Gu, C., and C. Li, "From Tumor Targeting to Speech Monitoring: Accurate Respiratory Monitoring Using Medical Continuous-Wave Radar Sensors," *IEEE Microwave Magazine*, Vol. 15, No. 4, 2014, pp. 66–76.

[34] van Loon, K., M. J. M. Breteler, L. van Wolfwinkel, et al., "Wireless Non-Invasive Continuous Respiratory Monitoring with FMCW Radar: A Clinical Validation Study," *Journal of Clinical Monitoring and Computing*, Vol. 30, No. 6, 2016, pp. 797–805.

[35] Li, C., J. Lin, and Y. Xiao. "Robust Overnight Monitoring of Human Vital Signs by a Non-Contact Respiration and Heartbeat Detector," in *EMBS'06, 28th Annual International Conference of the IEEE*, Engineering in Medicine and Biology Society, 2006, pp. 2235–2238.

[36] Santra, A., I. Nasr, and J. Kim, "Reinventing Radar: The Power of 4D Sensing, *Microw. J.*, Vol. 61, No. 12, 2018, pp. 26–38.

[37] Yardibi, T., P. Cuddihy, S. Genc, et al., "Gait Characterization via Pulse-Doppler Radar,"in *2011 IEEE International Conference on Pervasive Computing and Communications Workshops (PERCOM Workshops)*, IEEE, 2011, pp. 662–667.

[38] Kim, Y., and H. Ling, "Human Activity Classification Based on Micro-Doppler Signatures Using a Support Vector Machine," *IEEE Transactions on Geoscience and Remote Sensing*, Vol. 47, No. 5, 2009, pp. 1328–1337.

[39] Liu, L., M. Popescu, M. Skubic, M. Rantz, T. Yardibi, and P. Cuddihy, "Automatic Fall Detection Based on Doppler Radar Motion Signature," in *2011 5th International Conference on Pervasive Computing Technologies for Health care (Pervasive Health)*, IEEE, 2011, pp. 222–225.

[40] Shoup, D. C., "Cruising for Parking," *Transport Policy*, Vol. 13, No. 6, 2006, pp. 479–486.

[41] Werber, K., M. Rapp, J. Klappstein, M. Hahn, J. Dickmann, K. Dietmayer, and C. Waldschmidt, "Automotive Radar Gridmap Representations," in *2015 IEEE MTT-S International Conference on Microwaves for Intelligent Mobility (ICMIM)*, April 2015, pp. 1–4.

[42] Dube, R., M., Hahn, M. Schtz, J. Dickmann, and D. Gingras, "Detection of Parked Vehicles from a Radar Based Occupancy Grid," in *2014 IEEE Intelligent Vehicles Symposium Proceedings*, June 2014, pp. 1415–1420.

[43] Will, C., P. Vaishnav, A. Chakraborty and A. Santra, "Human Target Detection, Tracking, and Classification Using 24-GHz FMCW Radar," *IEEE Sensors Journal*, Vol. 19, No. 17, September 1, 2019, pp. 7283–7299.

[44] Shaker, G., S. Liu, and U. Wadhwa, "Non-Invasive Glucose Monitoring Utilizing Electromagnetic Waves," *MobileHCI*, Vienna, 2017.

[45] Diederichs, K., A. Qiu, and G. Shaker, "Wireless Biometric Individual Identification Utilizing Millimeter Waves." *IEEE Sensors Letters*, Vol. 1, No. 1, 2017, pp. 1–4.

2

Introduction to Deep Learning

Apart from the high-level integration offered by today's semiconductor technology, the other computing technology that has rapidly supported and enabled several applications of radar sensors is deep learning. Deep learning has grown and is now considered an independent field rather than a subset of machine learning. Broadly, there are two different types of learning algorithms: supervised learning and unsupervised learning. Supervised learning is a class of algorithm where during training a tuple of (X, y); that is, (input, labels) are presented and the task of the algorithm is to learn the mapping function such that $f : X \rightarrow y$. The objective of the algorithm is that during inference, when presented only X it is capable of predicting the correct label y. The supervised learning algorithm can be a classifier wherein the output variables are categorical; that is, good, bad, excellent, fair, and so on, or a regressor wherein the output variables have continuous values, like height of a person. Unsupervised learning is a class of algorithm where during training only input variable X is available, and the goal of the learning algorithm is to model underlying structures or distributions of the input data. Thus, unsupervised algorithms are used for clustering, where the objective is to find inherent groups available within the input data, such as finding a group of animals or birds from camera images, and association, where association rules are learned, such as the finding that beer is often bought together with diapers.

Machine learning algorithms are capable of extracting useful information from data by making use of hyperplanes in the feature domain of interest for a particular problem. Further, machine learning algorithms aim to reach a balance between bias, that is, approximating the in-sample data and variance, that is generalizing the out-sample data. On the other hand, through their architecture, deep learning algorithms exploit various structures and abstraction in the data by implicit feature learning instead of hand-crafted predesigned features. The algorithm refers to network architectures that contain multiple hidden layers at different depths and widths. Deep learning algorithms encapsulate a different hierarchy of representation, with the initial layers learning basic features and deeper layers learning more sophisticated and complex features from the data. Further, with the advent of deep learning and big data, the trade-off between bias and variance have been relaxed substantially and deep learning algorithms provide different mechanisms to handle problems of bias and variance. Deep learning algorithms have achieved state-of-the-art results in various computer vision, natural language processing, and speech processing tasks [1–3]. In computer vision problems such as image classification, image segmentation, object detections, image to text and text to image translation, image captioning, and super resolution, image/video synthesis has attained human-level or even better performance. However, the application of deep learning to radar data, and processing have been predominantly limited and far from the pace at which it has been successfully applied to computer vision problems. In this book, we aim to highlight and capture the deep learning applications to radar problems and their processing. We present the perspective from industry and academic research how deep learning has enabled several radar applications that were not considered feasible even a few years ago.

The chapter is laid out as follows. Section 2.1 introduces the perceptron and the different training methodologies proposed during the early deep learning era. Section 2.2 introduces the multilayer perceptron (MLP) and the important aspects associated with MLP training, including the bias-variance trade-off and curse of dimensionality. We also present the important activation functions and various optimizers used in MLP and also used in modern deep learning architectures. In Section 2.3, we present the convolutional neural network, its properties, and also describe the most popular CNN architectures in computer vision literature. Section 2.4 presents long short-term memory (LSTM) for learning temporal structures and dependencies in data. We introduce autoencoders and briefly present different types of autoencoder proposed in the literature in Section 2.5. In Section 2.6 and 2.7, we present the variational autoencoder (VAE), another special type of autoencoder, and generative adversarial network (GAN), an interesting and useful architecture involving both generative and discriminative models. Section 2.8 briefly presents robust deep learning, such as models that are robust to adverserial threats, and robustness metrics proposed in the literature.

2.1 Perceptron

Psychologist Frank Rosenblatt's idea of a perceptron [4] can be conceived as a simple mathematical model representation of how neurons in our brains work: a

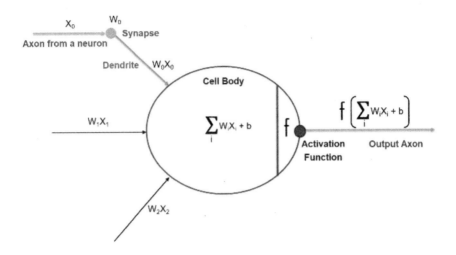

Figure 2.1 Rosenblatt's idea of a perceptron.

set of binary inputs (a nearby neuron's firing stage) is multiplied with continuous valued weight (the strength of the synapse of each connected nearby neuron) and thresholding is performed over the sum of these weighted inputs to give an output of 1 if the sum crosses the threshold or otherwise a 0 (meaning a neuron fires or not). A perceptron consists of a special bias input which ensures that more functions are computable with the same input by offsetting the summed value with a value of 1. Rosenblatt's idea of a perceptron is presented in Figure 2.1.

Inspired by the foundation work of Donald Hebb, who came up with the idea of how learning in the brain occurs through the formation and change of synapses (synaptic plasticity) between neurons, a learning algorithm for a perceptron was proposed as follows:

1. The weights of the perceptron are initialized randomly and a training dataset is defined;
2. For given inputs of an example from the training data set, the perceptron's output is computed;
3. If there is a mismatch between the perceptron output and the expected output, then
   - If the output should have been 0 but was 1, decrease the weights that had an input of 1
   - If the output should have been 1 but was 0, increase the weights that had an input of 1
4. Iteratively continue the process from 2–4, until the perceptron makes no mistakes.

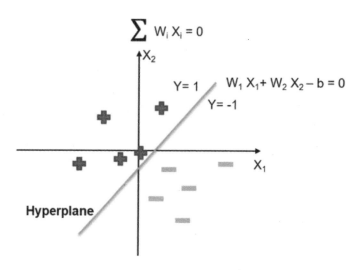

Figure 2.2   Perceptron hyperplane representation.

The output is the characteristic function of a half-space, which is limited by the hyperplane, as depicted in Figure 2.2.

Given a training set of T examples as tuples of (N coordinates/features, class) that is, $((x_1^j, x_2^j, \cdots x_N^j), t^j)$ where $t^j \in \pm 1$ $\quad \forall j = 1, 2 \cdots T$. The objective of the perceptron training algorithm is to find the weights w_i such that

$$t^j = \text{sign} \left(\sum_{i=1}^{N} w_i x_i^j \right) \quad \forall j = 1, 2, \cdots T \tag{2.1}$$

The training algorithm is iterated as follows;

- Initialize weights to random values $w_i(0)$
- For each training example j:
  - Compute the output $y^j(k)$ for the kth iteration as

$$y^j(k) = \text{sign} \left(\sum_{i=0}^{N} w_i(k) x_i^j \right) \tag{2.2}$$

  - If $t^j \neq y^j(k)$ then update the weights $w_i(k)$ for $k + 1$ iteration as

$$w_i(k + 1) = w_i(k) + \left(t^j - y^j(k) \right) x_i^j \tag{2.3}$$

The process is continued until we achieve correct predictions for all of the examples in the training data set. However, it must be noted that this algorithm

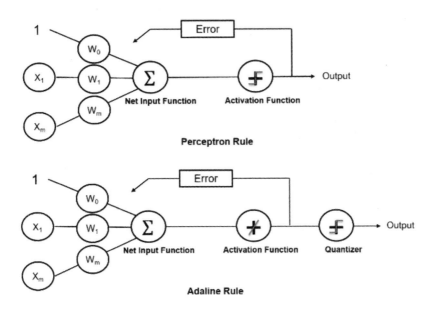

Figure 2.3 Perceptron rule vs. adaline rule.

would converge only for linearly separable cases. A few years after the discovery of perceptron, Bernard Widrow and his doctoral student Tedd Hoff proposed an adaptive linear neuron (Adaline) [5]. Unlike in the perceptron, the weights are updated based on a linear activation function $g(z)$, which is just an identity function of the net input rather than a unit step function. A comparison of the unit step activation function in a perceptron to the linear activation function is presented in Figure 2.3.

The biggest advantage of the linear activation function over the unit step function is that it is continuous and thus differentiable. This property allows us to define a cost function $J(w)$ similar to ordinary least squares linear regression as the sum of squared errors and minimize it with a simple yet very useful optimization technique called gradient descent.

The cost function can be defined as

$$J(w) = \frac{1}{2} \sum_{i=1}^{T} \left(t^{(i)} - o^{(i)} \right)^2 \quad \forall o^{(i)} \in \mathbb{R} \tag{2.4}$$

The principle behind gradient descent is that a step is taken in the opposite direction of the gradient and the step size is dependent on the learning rate and the slope of the gradient, as visualized in Figure 2.4.

In the adaline learning rule, each update takes place by taking a step in the opposite direction of the gradient and thus we need to compute the partial

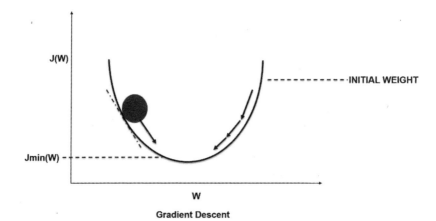

Figure 2.4 Gradient descent algorithm.

derivative for each weight of the cost function.

$$\Delta w_j = -\eta \frac{\delta J}{\delta w_j} \tag{2.5}$$

The partial derivative of the cost function with respect to a jth weight can be calculated as follows:

$$\frac{\delta J}{\delta w_j} = \frac{\delta}{\delta w_j} \frac{1}{2} \sum_i (t^{(i)} - o^{(i)})^2$$

$$= \frac{1}{2} \sum_i \frac{\delta}{\delta w_j} (t^{(i)} - o^{(i)})^2$$

$$= \frac{1}{2} \sum_i 2(t^{(i)} - o^{(i)}) \frac{\delta}{\delta w_j} (t^{(i)} - o^{(i)})$$

$$= \sum_i (t^{(i)} - o^{(i)}) \frac{\delta}{\delta w_j} \left(t^{(i)} - \sum_j w_j x_j^{(i)} \right)$$

$$= \sum_i (t^{(i)} - o^{(i)})(-x_j^{(i)}) \tag{2.6}$$

where (t, o) represents the target and output, respectively.

And thus we can write the weight updates as

$$\Delta w_j = -\eta \frac{\delta J}{\delta w_j} = -\eta \sum_i (t^{(i)} - o^{(i)})(-x_j^{(i)})$$

$$= \eta \sum_i (t^{(i)} - o^{(i)}) x_j^{(i)} \tag{2.7}$$

Finally we can use the weight update as in perceptron training algorithm. In this learning algorithm the weight update is calculated over all examples in the training data set. Since the cost function is computed over all examples, the computation cost is high as the examples increase; however, the cost function is always monotonically nonincreasing; that is, $J(w_j(k + 1)) \leq J(w_j(k))$. In the case of the convex cost function, the training would converge to the global minimum if the step size is set appropriately. However, in the case of a nonconvex cost function, depending on the initial start the training step could be stuck in local minima.

Two alternate approaches are to consider one example per iteration, which would require low computation cost but then the nonincreasing property of the cost function with iterations are lost. Alternately, a set of examples are taken for updating the gradient at each iteration, which is refered as batch gradient descent. This is computationally the simplest, however, the number of epochs or iterations required to converge would increase drastically. Figure 2.5 illustrates the gradient steps of the stochastic gradient descent [6] and batch gradient descent algorithm over the cost function with the two weights, w_1, w_2. Table 2.1 presents the summary of properties of the perceptron training algorithm compared to Adaline.

2.2 Multilayer Perceptron

In multi-layer perceptron (MLP) or also called feed forward neural network (FFNN), there can be more than a single linear layer or neuron. A simple example of MLP consists of an input layer, a hidden layer in middle, and an output layer. The number of hidden layers can be increased to increase the complexity of the model. In MLP, the output of each neuron in a layer $i - 1$ is fed as input to all the neurons in layer i as depicted in Figure 2.6.

For example, MLP finds the best approximation for a classifier $y = f(x; \theta)$ that maps input $x \in X$, where X is the set of training examples, to an output class y

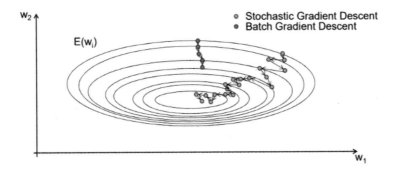

Figure 2.5 Comparison of stochastic gradient descent and batch gradient descent algorithm.

Table 2.1
Comparison of Perceptron Training to Adaline

Perceptron Training	Adaline
Unit step activation function	Linear activation function
Simple training algorithm	Global/batch/stochastic gradient Descent training algorithm
No parameters	parameters : learning rate/step size
Converges in finite number of iterations	converges asymptotically
Works only for linearly seperable problems	Works for nonlinear problems but would suffer misclassifications

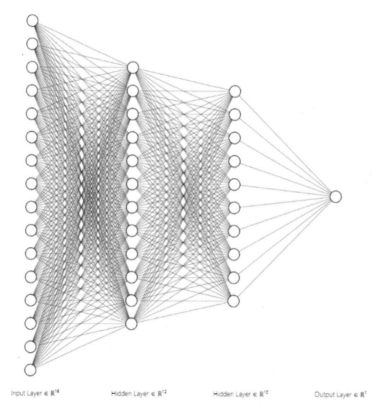

Input Layer $\in \mathbb{R}^{16}$ Hidden Layer $\in \mathbb{R}^{12}$ Hidden Layer $\in \mathbb{R}^{10}$ Output Layer $\in \mathbb{R}^{1}$

Figure 2.6   Multilayer perceptron.

and learns the best set of parameters θ. Each layer is represented as $y = f(Wx + b)$ where f is the activation function, W is the set of weights, x is the input vector that also can be the output of the previous layer, and b is the bias vector. A popular strategy to train MLP is to initialize the weights randomly and then change them iteratively to achieve a lower loss and the change is done using a gradient descent optimization technique.

2.2.1 Training

For MLP training, there are two steps: namely forward pass and error propagation.

1. *Forward pass*: In this stage, we feed the input to the model and multiply with weight vectors and add bias for each layer to compute the output of the model. Table 2.2 depicts the input, activation, and output at each layer of the MLP calculation.

2. *Error backpropagation*: Consider one sample for which inputs $(x_1, x_2, \cdots, x_n)$ and expected outputs $(t_1, t_2, \cdots t_k \cdots t_m)$ and real outputs $(y_1, y_2, \cdots y_k, \cdots y_m)$. The error for one sample is therefore $E = \frac{1}{2} \sum_{k=1}^{m} (y_k - t_k)^2$ where y_k is a function of the weights $w_{ij}^{(l)}$. The weights are updated as follows to minimize the error using the gradient descent algorithm as

$$w_{ij}^{(l)} \leftarrow w_{ij}^{(l)} - \lambda \frac{\partial E}{\partial w_{ij}^{(l)}} \tag{2.8}$$

In (2.8), $\frac{\partial E}{\partial w_{ij}^{(l)}}$ can be computed as the following steps:

(a)

$$\frac{\partial E}{\partial w_{ij}^{(l)}} = \frac{1}{2} \sum_k \frac{\partial E}{\partial y_k} \frac{\partial y_k}{\partial w_{ij}^{(l)}} \tag{2.9}$$

where $\frac{\partial E}{\partial y_k} = (y_k - t_k)$

(b) Since y_k is a function of $u_j^{(l)}$ only then

$$\frac{\partial y_k}{\partial w_{ij}^{(l)}} = \frac{\partial y_k}{\partial u_j^{(l)}} \frac{\partial u_j^{(l)}}{\partial w_{ij}^{(l)}} \tag{2.10}$$

Table 2.2
Layerwise Input, Activation, and Output Notations

operations	Layer 1	Layer 2	$\cdots$	Layer L
inputs	$x_i^{(1)} = x_i$	$x_i^{(2)} = o_j(1)$	$\cdots$	$x_i^{(l)} = o_j(L-1)$
activation	$u_j^{(1)} = \sum_{i=1}^{N} w_{ij}^{(1)} x_i^{(1)}$	$u_j^{(2)} = \sum_{i=1}^{N} w_{ij}^{(2)} x_i^{(2)}$	$\cdots$	$u_j^{(L)} = \sum_{i=1}^{N} w_{ij}^{(L)} x_i^{(L)}$
output	$o_j^{(1)} = f(u_j^{(1)})$	$o_j^{(2)} = f(u_j^{(2)})$	$\cdots$	$o_j^{(L)} = f(u_j^{(L)})$

(c)

$$\frac{\partial u_j^{(l)}}{\partial w_{ij}^{(l)}} = o_i^{(l-1)} \qquad (2.11)$$

which is computed during the feed forward step.

Thus putting it all together,

$$\frac{\partial E}{\partial w_{ij}^{(l)}} = (y_j - t_j)\frac{\partial y_k}{\partial u_j^{(l)}}o_i^{(l-1)} = \frac{\partial E}{\partial u_j^{(l)}}o_i^{(l-1)} \qquad (2.12)$$

Some important aspects during MLP training are

1. *Learning rate*: Each weight update is controlled by parameter λ, which is known as the learning rate parameter. If the learning rate is too small then it may result in very slow learning, can get trapped in local minima easily, and can keep running for many iterations. On the other hand, if the learning rate is large then it may step over the minima, and can fail to converge and potentially diverge. Therefore, it is really important to choose a good learning rate based on the architecture, data set, transfer function, and so on. Figure 2.7 illustrates the effects of choosing small and large learning rates on the gradient descent.

2. *Weight initialization*: It is important to randomize the weights during initialization, otherwise symmetry in weights would prevent the network from learning. Usually small random values are used, which is highly important when the number of neuron in a layer grows, as the weighted sum may saturate the transfer function.

3. *Overfitting and Underfitting*: In machine learning, the objective is not only to minimize the cost function on in-sample data (i.e., data available or seen), but also generalize on out-sample data (i.e., data not available or unseen). During training, the available data set is divided into training data set, validation data set, and test data set. The training data set is used to train the model, validation data set is used to set the hyperparameters of the model, and test data set is used for estimating the out-sample or generalization accuracy.

   When the performance is poor on the training data then it can be regarded as underfitting and is often due to poor choice of learning rate or if the NN is under-dimensioned. This error is refered to as 'bias'. The issue of underfitting is illustrated in Figure 2.8. The issue of overfitting arises when performance is good on the training data (i.e., good approximation accuracy), but poor on the test/validation data; in other words, poor generalization accuracy. This phenomenon is also refered as 'variance' and is illustrated in Figure 2.8. If the training set

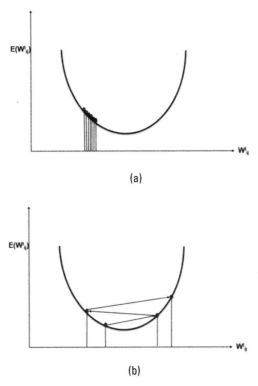

(a)

(b)

Figure 2.7 Illustration of gradient descent when (a) learning rate is small, and (b) learning rate is large.

size is insufficient or the model complexity is high for the data, the model memorizes or approximates the training data well but does not generalize well on test data (i.e., overfits). The purpose of training a machine learning model is to find a model as shown in Figure 2.8 so that both training accuracy (bias) and generalization accuracy (variance) are minimized. Typically for MLP models, the process of training finds a model so that a balance between bias and variance can be achieved and often is referred to as a bias-variance trade-off. As we would also see later, in the case of deep learning the bias-variance trade-off is not applicable since there are seperate mechanisms to reduce bias and variance, thus the trade-off is not readily applicable.

4. *Curse of dimensionality:* The other critical aspect in machine learning in general is the curse of dimensionality. The curse of dimensionality is closely related to overfitting. In high-dimensional spaces, most of the training data reside in the corners of the hypercube defining the feature space. Instances in the corners of the feature space are much more difficult to classify than instances around the centroid of the

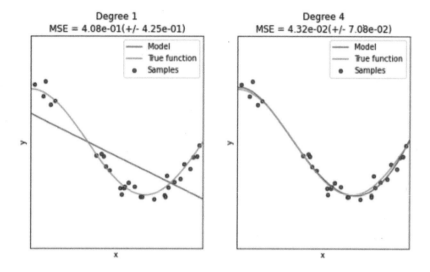

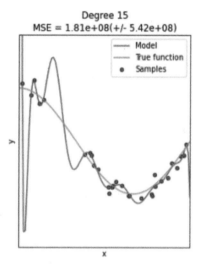

Figure 2.8　Illustration of underfitting and overfitting of model.

hypersphere. Thus, as the number of feature or dimensions grows, the amount of data we need such that the model generalizes accurately also grows exponentially.

2.2.2　Activation Functions

Some of the standard activation functions are as follows:

1. *Sigmoid (logistic activation)*: This activation function is originally inspired by the real neuron. The output of this activation function is between

[0, 1]. Its major drawbacks are that a saturated neuron will not learn and the activation is computationally expensive.

$$f(x) = \frac{1}{1 + \exp^{-x}}$$

$$f'(x) = f(x)(1 - f(x)) \tag{2.13}$$

2. *Hyperbolic tangent activation*: The output range of hyperbolic tangent activation is between $[-1, 1]$ and is zero centered. Like sigmoid activation, this activation also does not train saturated neurons.

$$f(x) = \tanh(x) = \frac{1 - \exp^{-x}}{1 + \exp^{-x}}$$

$$f'(x) = 1 - \tanh^2(x) \tag{2.14}$$

3. *Rectified linear unit (ReLu) [7]*: In ReLu activation, there is no saturation if $x > 0$ and it is more computationally efficient and leads to faster convergence. In such an activation the output is always positive and the inactive neurons do not learn.

$$f(x) = \max(0, x)$$

$$f'(x) = \begin{cases} 1 & \text{if } x > 0 \\ 0 & \text{if } x < 0 \end{cases} \tag{2.15}$$

4. *Leaky rectified linear unit [8] and parametric ReLu [9]*: These activation functions are an improvement over the normal ReLu, which overcomes the problem of dead neurons. In the case of a leaky ReLu α is a small constant, such as 0.01. In the case of a parametric ReLu α is a hyperparameter learned through backpropagation.

$$f(x) = \begin{cases} x & \text{if } x > 0 \\ \alpha x & \text{if } x < 0 \end{cases}$$

$$f'(x) = \begin{cases} 1 & \text{if } x > 0 \\ \alpha & \text{if } x < 0 \end{cases} \tag{2.16}$$

Figure 2.9 presents the commonly used activation functions described above, namely, sigmoid function, hyperbolic function, ReLu and leaky rectified linear unit.

One of the other standard activation functions, typically used in the output layer for classification problems, is a softmax layer. Since the squared error is not suitable for such cases where classes are mutually exclusive, a better approach is to

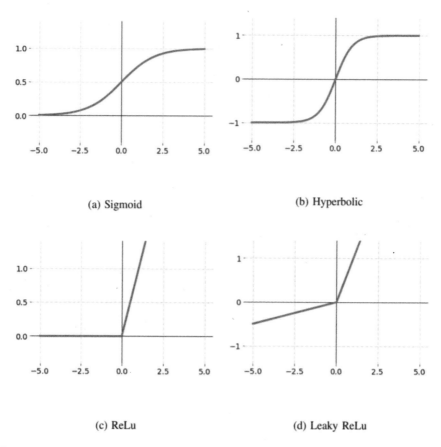

Figure 2.9 Illustration of various activation functions: (a) sigmoid function, (b) hyperbolic function, (c) ReLu function and (d) leaky ReLu.

assign probabilities to each class with the constraint that the outputs should sum up to 1. The softmax function [10] forces the output to represent a probability distribution across the possible classes, L. Its function and its derivative are given as follows:

$$p_k = \frac{\exp^{x_k}}{\sum_{j=1}^{L} \exp^{x_j}}$$

$$p'_k = f(x_k)(1 - f(x_k)) \tag{2.17}$$

Figure 2.10 presents the softmax layer used at the output layer for classification. The cost function typically associated with the softmax layer is the negative log probability of the correct prediction, called cross-entropy or

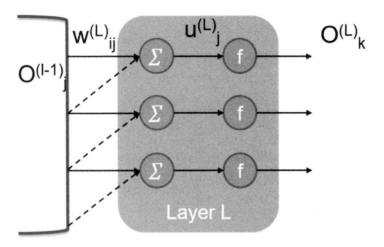

Figure 2.10 Softmax layer.

log-loss cost function and is defined as

$$E(t, p) = -\sum_{k=1}^{L} t_k \log \left(\frac{\exp^{u_k}}{\sum_{j=1}^{L} \exp^{u_j}} \right) \tag{2.18}$$

2.2.3 Optimizers

Various modifications have been proposed as an advancement over the standard stochastic gradient descent algorithm to improve training of neural networks and are described below.

1. *Momentum*: It accelerates the SGD toward the relevant direction while reducing the oscillations. It basically adds part of the previous weight updates to the current update vector.

$$v_t = \gamma v_{t-1} + \lambda \delta_w E(w) \quad w_{t+1} = w_t - v_t \tag{2.19}$$

2. *Nesterov accelerated gradient [11]*: While accelerating convergence by momentum an error is introduced. This is countered in the Nesterov accelerated gradient by including part of the previous weight updates to the current update vector to perform the weight update.

$$v_t = \gamma v_{t-1} + \lambda \delta_w E(w - \gamma v_{t-1}) \quad w_{t+1} = w_t - V_t \tag{2.20}$$

Typical values of $\gamma = 0.9$.

3. *Adagrad [12]*: The motivation of Adagrad is to have an adaptive learning rate for each parameter. Large updates should occur for infrequent parameters and smaller updates for frequent parameters.

For each weight update, the learning rate is adapted as

$$w_i^t = w_i^{t-1} - \frac{\lambda}{\sqrt{G_i^{t-1} + \epsilon \delta_w E(w_i^{t-1})}}$$

$$G_i^{t-1} = \sum_{tt=0}^{t-1} (\delta_w E(W_i^{tt}))^2 \tag{2.21}$$

since the sum of the squared gradient grows continuously, it may lead to a very small learning rate.

4. *RMSprop [13]*: RMSprop is an improvement over the Adagrad where learning continues after many parameter updates. A moving average of a squared gradient for each weight is computed by limiting the gradient accumulation to a certain past.

$$w_i^t = w_i^{t-1} - \frac{\lambda}{\sqrt{G_i^{t-1} + \epsilon \delta_w E(w_i^{t-1})}}$$

$$G_i^t = \gamma G_i^t + (1 - \gamma)(\delta_w E(W_i^{t-1}))^2 \tag{2.22}$$

5. *Adadelta [14]*: Adadelta is another improvement over the Adagrad to continue learning after many parameter updates. Here the gradient accumulation is limited to a certain past update by computing a moving average of both the squared gradient and parameter updates for each weight parameter as

$$w_i^t = w_i^{t-1} - \lambda^t \delta_w E(w_i^{t-1}) = w_i^{t-1} + v_i^{t-1}$$

$$G_i^t = \gamma G_i^{t-1} + (1 - \gamma)(\delta_w E(w_i^{t-1}))^2$$

$$\Delta_{wi}^t = \gamma \Delta_{wi}^{t-1} + (1 - \gamma)(v_i^t)^2$$

$$\lambda^t = \frac{\sqrt{\Delta_{wi}^{t-1} + \epsilon}}{\sqrt{G_i^t + \epsilon}} \tag{2.23}$$

6. *Adaptive moment estimation (Adam) [15]*: This adds momentum to RMSprop by using first and second moments (mean and variance) of the gradient.

$$m_i^t = \beta_1 m_i^{t-1} + (1 - \beta_1)\delta_w E(w_i^{t-1})$$

$$v_i^t = \beta_2 v_i^{t-1} + (1 - \beta_2)(\delta_w E(w_i^{t-1}))^2$$

$$w_i^t = w_i^{t-1} - \frac{\lambda}{\sqrt{v_i^t + \epsilon}} m_i^t \tag{2.24}$$

2.2.4 Types of Models

There are broadly two types of classifiers in the context of supervised learning: generative models and discriminative models. A generative algorithm models how the data is actually generated and learns the joint probability distribution $P(X, Y)$, where X is the input features and Y is the corresponding classification label. A discriminative algorithm models how to classify input features, without providing a model of how the features are generated, by learning to maximize the conditional probability distribution $P(Y|X)$. A discriminative model models the decision boundary between the classes, whereas a generative model explicitly models the actual distribution of each class. An example of a generative model is naive Bayes classifier and a discriminative model is logistic regression.

The generative model predicts the conditional probability required for classification with the help of Bayes theorem, while the discriminative model learns the conditional probability distribution without prior assumption. Thus while both models obtain classification, they learn different distributions. Generative models require some structural assumptions on the model, while discriminative models don't. For example, naive Bayes assumes conditional independence of input features, while logistic regression does not require such assumptions.

Generative models often outperform discriminative models on smaller data sets because the generative model uses some structural assumptions on the model that acts as regularization and prevents overfitting. However, the discriminative model tends to outperform its generative counterpart as the data starts to grows.

2.3 Convolutional Neural Networks

In deep learning, the most widely used network are CNNs [16] for tasks such as object detection, face recognition, image segmentation, and super-resolution. In CNNs, the image classification is performed by incorporating various layers, namely convolution layers, pooling layers, and dense layers with softmax loss. The network sees an image as a multidimensional array of pixels and based on the resolution of the image $h \times w \times d$ (h = Height, w = Width, d = Dimension). For example, an image of $32 \times 32 \times 3$ represents a red green blue (RGB) image (3-dimensional) while an image of $32 \times 32 \times 1$ is a gray-scale image.

2.3.1 Convolution Layer

The first spatial feature extraction in a CNN is done by convolution layers. In convolution layers, we compute the output of a dot product between a filter or kernel (weight matrices) and areas based on the size of the filter of the input image. The filter slides through the entire image performing the same dot product that results in an intermediate output matrix. Note that the filter must have the same number of channels as in the input image.

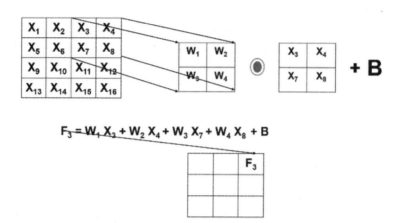

$$F_3 = W_1 X_3 + W_2 X_4 + W_3 X_7 + W_4 X_8 + B$$

Figure 2.11   Illustration of convolution operation.

The output dimension of the convolution layers are calculated as follows:

$$w_{out} = \frac{w_{in} - f_w + 2p}{s} + 1$$

$$h_{out} = \frac{h_{in} - f_h + 2p}{s} + 1 \qquad (2.25)$$

where w_{in}, h_{in} are the width and the height of the input image, f_w, f_h are the width and height of the filter or kernel, p, s are the padding and the stride factors, and are set ≥ 1. Figure 2.11 illustrates the convolution operation. The weights of the filters are learnt by the network through backpropagation. Various kernel types are presented in Figure 2.12.

The components of the convolutional layers are described as follows:

1. *Strides*: While applying convolution, we make shifts to move the filter across the entire image. Stride defines the degree of shift; for example, if the stride is 1 then the filter is shifted by a single pixel and if 2 then it is shifted by two pixels and so on. Figure 2.13 presents the stride of two applied to the incoming image.

2. *Padding*: When a size-defined filter does not fit the input image properly, then we have two options:
   - *Zero-padding*: Pad the image with zeros so that the filter fits perfectly
   - *Valid-padding*: Dropping the parts of images that do not fit the filter perfectly

3. *Activation function*: ReLu is mostly used as activation function for convolution layers to introduce nonlinearity in our convolution

OPERATION	FILTER	CONVOLVED IMAGE
IDENTITY	$\begin{pmatrix} 0\ 0\ 0 \\ 0\ 1\ 0 \\ 0\ 0\ 0 \end{pmatrix}$	
EDGE DETECTION	$\begin{pmatrix} 1\ 0\ \text{-}1 \\ 0\ 1\ 0 \\ \text{-}1\ 0\ 1 \end{pmatrix}$	
EDGE DETECTION	$\begin{pmatrix} 0\ 1\ 0 \\ 1\ \text{-}4\ 1 \\ 0\ 1\ 0 \end{pmatrix}$	
EDGE DETECTION	$\begin{pmatrix} 0\ 0\ 0 \\ 0\ 1\ 0 \\ 0\ 0\ 0 \end{pmatrix}$	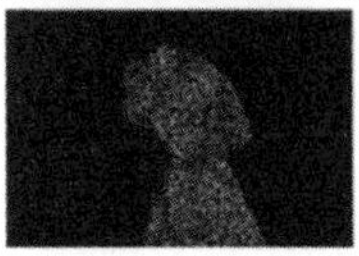
SHARPEN	$\begin{pmatrix} 0\ \text{-}1\ 0 \\ \text{-}1\ 5\ \text{-}1 \\ 0\ \text{-}1\ 0 \end{pmatrix}$	
BOX BLUR	$1/9\begin{pmatrix} 1\ 1\ 1 \\ 1\ 1\ 1 \\ 1\ 1\ 1 \end{pmatrix}$	

Figure 2.12   Illustration of various kernel filters and their effects on the input image.

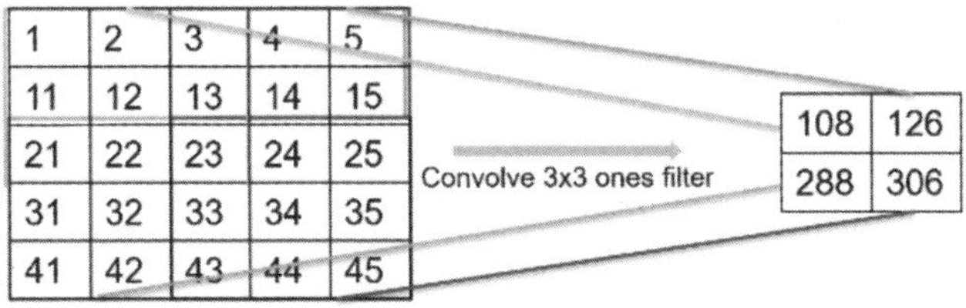

Figure 2.13   Illustration of stride of two on the incoming image data.

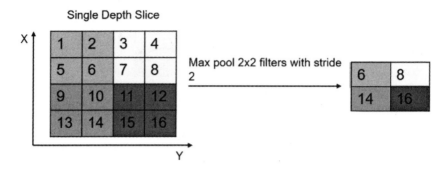

Figure 2.14   Illustration of max-pooling operation after convolution layer.

neural network. Further, ReLu being a simple function improves the backpropagation iteration. However, depending on the use case, one may opt for other activation functions.

4. *Pooling layers*: Pooling layers or more specifically spatial pooling layers perform subsampling or downsampling of the input image while retaining the most relevant information. This process helps in the reduction of parameters when the images are too large. The most commonly used pooling layers are

- *Max pooling*: Taking the largest element within the defined nonparametric filter size. Figure 2.14 illustrates the max-pooling operation.

- *Average pooling*: Taking the average of all the elements within the defined nonparametric filter size.

- *Sum pooling*: Taking the sum of all the elements within the defined nonparametric filter size.

5. *Dense layers*: At the end of the convolution neural network, a single dense layer or multiple dense layers are used to which the flattened (1D array) of the outputs of convolutional and pooling layers are fed with the activation function as sigmoid or softmax to classify the outputs.

Figure 2.15 depicts an example of a convolutional neural network applied on an input image.

2.3.2 Popular Architectures

2.3.2.1 LeNet-5
In 1998, Lecun [17] proposed a convolutional neural network for the purpose of automatic classification of handwritten digits on bank checks. This network is a convolutional neural network containing local receptive fields, spatial

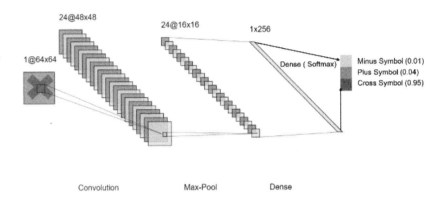

Figure 2.15 Convolutional neural network.

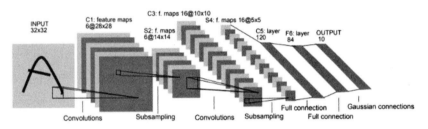

Figure 2.16 LeNet-5 architecture for handwriting recognition.

subsampling, and shared weights. This was the first time that digit recognition was done without handcrafted feature engineering. The network is composed of seven layers, which include three convolution layers of filter size 5×5 and a stride of 1 and 2 subsampling layers performing 2×2 average pooling. The subsampling layers are after the first two convolution layers, respectively, for the purpose of reducing the number of trainable parameters as computationally limitation was a major hurdle at that point in time.

The proposed architecture achieved classification accuracy of 95% on the test data. The output layer is a fully connected layer with softmax activation yielding a probabilistic distribution over the possible digit classes. Recent implementation of LeNet-5 uses max-pooling instead of average-pooling as it picks up the highest value in a 2×2 grid, which helps in having bigger gradients for backpropagation and speeding up the training process. The LeNet-5 architecture is presented in Figure 2.16.

2.3.2.2 SqueezeNet

SqueezeNet architecture [18] was proposed by DeepScale, UC Berkeley, and Stanford University, which offers smaller CNN architectures with matching accuracy to those of the large networks such as AlexNet [19]. With regard to building embedded solutions, smaller CNNs are necessary due to memory and computational power limitation in embedded systems. The main architectural ideas behind SqueezeNet are as follows:

1. *Replacing 3 × 3 filters with 1 × 1 filters*: Replacing the majority 3 × 3 filters with pointwise 1 × 1 filters leads to nine times less parameters than the use of 3 × 3 filters.
2. *Using a squeeze layer to reduce depth of input to the 3 × 3 filters*: The number of input channels to the 3 × 3 filters are reduced by the use of 1 × 1 filters as a bottleneck.
3. *Late downsampling*: The downsampling is done much later in the network to preserve large feature maps to maximize accuracy.

The building block of SqueezeNet is the fire module, which comprises a 1 × 1 convolution filter layer called the squeeze layer and an expand layer consisting of a mix of 3 × 3 and 1 × 1 convolution filter layers as shown in Figure 2.17.

The squeeze layer does not extract any spatial information but helps to reduce the number of channels, thus lowering the computation to be done by the 3 × 3 filters. The architecture of SqueezeNet includes a convolution layer in the beginning followed by eight fire modules with a gradual increment in the number of filters in them and ends with a final convolution layer. Max-pooling is performed after the convolution layers and third and seventh fire module with a stride of 2. Figure 2.18 presents the SquezeNet architecture that demonstrated the same accuracy as that of Alexnet with a 50 times smaller model size on ImageNet data set.

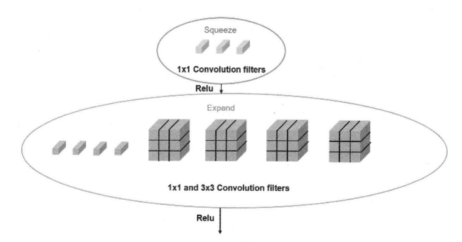

Figure 2.17   Squeeze and expand filters used in SqueezeNet.

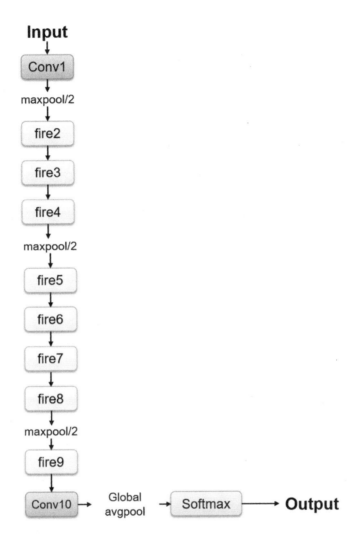

Figure 2.18   SqueezeNet architecture.

2.3.2.3   Inception Network

Inception Network v1 was initially proposed by the authors in their paper [20] with an inception module as the major building block. The main intuition behind the inception block is the use of multiple-sized filters at the same level to make the layer wider in nature and more flexible to images where the amount of pixels covered by the object varies. The inception module is depicted in Figure 2.19.

The different blocks allows to perform convolution with different filter sizes on the same input. The 1×1 convolutions used before the 3×3 and 5×5 reduce the number of channels before feeding it to the two larger convolution filters to

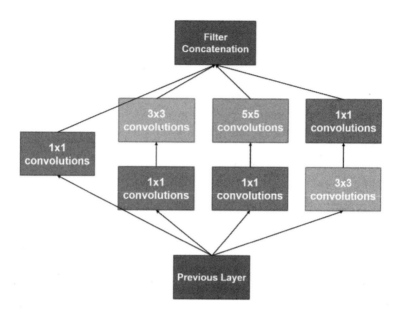

Figure 2.19   Inception module.

reduce computations. It also includes a max-pooling layer with 3×3 filter size and a 1×1 convolution layer on top of it to similarly reduce the channels. The final output of the module is a concatenation of the output of these filters.

Inception v1 uses nine such inception module stacked one after another. Unlike most other networks, it uses a global average-pooling layer instead of a fully connected (FC) layer in the end. This reduces the model size even more and allows the network to be less prone to overfitting. It also uses two auxiliary losses that are added to the real loss with a weight of 0.3 for each. The authors apply softmax on the output of the two subnetworks and computes auxiliary loss over the same labels, which acts as regularizer and prevents a vanishing gradient problem. These auxiliary classifiers are used for training purpose only. The inception network has evolved over the years and the most-used form of inception networks is Inception v3. The Inception v3 network [21] has the following major upgrades over v1:

1. *Factorized convolutions*: The use of factorized convolutions allows reduction in the number of parameters with boost in accuracy. For example, 5×5 filters can be replaced by two 2×2 convolutions. The network also uses an asymmetric factorized convolution by replacing $n \times n$ filters with $1 \times n$ and $n \times 1$ filters.

2. *Auxiliary classifier as regularizer*: There is only one auxiliary classifier in inception v3 and it is used as a regularizer unlike in v1 with two auxiliary classifiers for preventing a vanishing gradient problem.

3. *Label smoothing*: Label smoothing is done to prevent overfitting by not allowing some labels to become larger than others. This is necessary in scenarios where the number of classes is quite large. The new labels are computed recursively as $(1 - \epsilon)$old labels $+ \epsilon/K$, where ϵ is a hyperparameter and K denotes the number of classes.

2.3.2.4 Residual Network

Deep convolution networks achieve very high accuracy in tasks such as image classification by stacking multiple layers. However, in many experiments it was observed that more and more layers yielded poor results compared to the shallow networks. This should not be the case as neural networks are designed to approximate complex functions and so even if the extra layers do not learn new features; they should have been able to model an identity function and act as skip layers. In the real world, due to vanishing gradient and curse of dimensionality, this is not achieved while using deeper networks. In the ResNet [22] paper, the authors introduce residual block, which explicitly forces the network to model identity mapping. The residual is the difference $F(x)$ between the input x to the subnetwork and actual output $H(x)$ of it. This provides the network the ability to skip a subnetwork by setting $F(x) = 0$. In backpropagation, learning of the residual gives the network the option to forward gradient from higher layers directly to lower layers ignoring layers in between. A typical residual block is depicted in Figure 2.20.

There are multiple variants of residual blocks proposed in the literature and these are presented in Figure 2.21. Based on the problem and data set, one variant

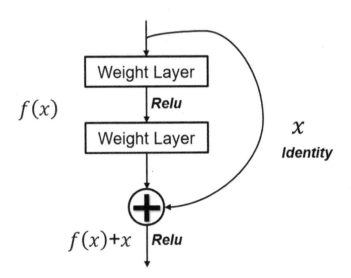

Figure 2.20 Residual block.

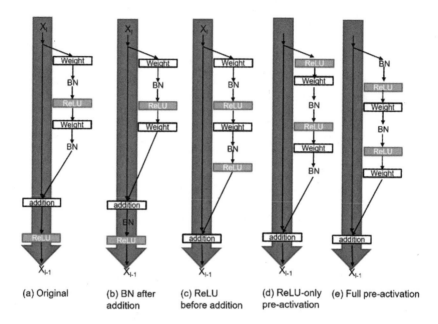

(a) Original	(b) BN after addition	(c) ReLU before addition	(d) ReLU-only pre-activation	(e) Full pre-activation

Figure 2.21   Residual block variations.

can perform better than the other. However, mostly in practice, the best results are observed using a full preactivation residual block.

2.3.2.5 Efficient Network

In May 2019, the authors introduced the idea of EfficientNets in their seminar paper [23]. The paper introduced the idea of scaling CNNs efficiently, as the name suggests. Model scaling involves the process of scaling an existing model in terms of depth, width, or resolution. The most popular scaling is depth scaling and in general this manual scaling process yields better performance. However, after a certain extent of depth scaling there is no more improvement in performance and the model begins to be adversely affected. Wide scaling involves having smaller networks by having shallow yet wider networks that tend to capture more fine-grained features. Resolution scaling is built on the idea of having a high-resolution input that will have more fine-grained information and thus increase the accuracy. However, in both wide scaling and resolution scaling, accuracy seems to saturate after a certain degree of scaling. The authors introduced the idea of compound scaling, which involves scaling strategically in all three attributes for delivering better results.

Figure 2.22 presents the various scalings to improve the performance of a neural network. The compound scaling process involves the use of a compound user-specific coefficient ϕ to scale in depth, width, and resolution together. EfficientNet proposes to obtain the depth, width, and resolution scaling for a

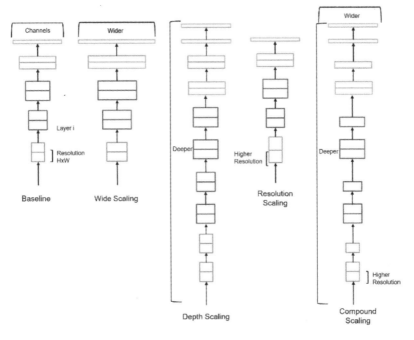

Figure 2.22   Different types of scaling: (a) baseline, (b) wide scaling, (c) depth scaling, (d) resolution scaling, and (e) compound scaling.

network using the following equation:

$$\text{depth} : d = \alpha^\phi$$

$$\text{width} : w = \beta^\phi$$

$$\text{resolution} : r = \gamma^\phi$$

$$s.t. \ \alpha\beta^2\gamma^2 \approx 2$$

$$\alpha \geq 1, \beta \geq 1, \gamma \geq 1 \tag{2.26}$$

where α, β, and γ are the scaling factor obtained for depth, width, and resolution scaling, respectively, through a grid search approach. The authors have reported significant increment in performance when compound scaling is performed with famous baseline models such as MobileNet and ResNet-50. Since the overall performance of the scaling is very much dependent on the baseline model, the authors proposed a new baseline model called EfficientNet B0 that was developed by using neural architecture search (NAS) that optimizes on both accuracy as well as floating point operations needed. The EfficientNets produced state-of-the-art accuracy on ImageNet while being very efficient compared to its other networks. The architecture of the proposed B0 model is provided in Table 2.3.

Table 2.3
Architecture of Proposed B0 EfficientNet in [23].

Stage (i)	Operation Operation	Resolution ($H_i \times W_i$)	Number of Channels (C_i)	Number of Layers (L_i)
1	Conv, 3x3	224x224	32	1
2	MBConv1, k3x3	112x112	16	1
3	MBConv6, k3x3	112x112	24	2
4	MBConv6, k5x5	56x56	40	2
5	MBConv6, k3x3	28x28	80	3
6	MBConv6, k5x5	14x14	112	3
7	MBConv6, k5x5	14x14	192	4
8	MBConv6, k3x3	7x7	320	1
9	Conv1x1 and Pooling and FC	7x7	1280	1

The 7 MBConv blocks comprise inverted residual blocks where the skip connections connect narrow layers and the wide layers between skip connections. It also houses a squeeze and excite block. The huge reduction in computational power and memory requirements despite scaling makes EfficientNets highly mobile and on-the-edge friendly.

2.3.3 Transfer Learning

In practice, pretrained CNN architectures are preferred over training a CNN architecture from scratch. This is because of the unavailability of large training data that often leads to poor training and overfitting. The process of using a pretrained network to solve another task with a different data set is known as transfer learning. For example, one of the most popular image classification networks called ImageNet, which is trained on millions of labeled images, is often used as a pre-trained network for other object-specific classification task, due to its robust and powerful feature extraction capabilities in the case of images. Transfer learning [24] is not restricted to just images; in fact this methodology is a common practice in all domains like speech recognition, text classification, and much more.

Transfer learning can be used for the following purposes:

1. The pretrained network can act as a robust feature extractor for the given data set, where the output of intermediate layer, which is often the last layer before the classification layer, is regarded as having features that can be fed as input to train a linear classifier or support vector machine (SVM).

2. Instead of using random initialization, the pretrained network can also be used to initialize the weights of a similar network that is to be trained on a different data set to speed up the training process.

3. Fine-tuning is the most common practice in transfer learning where a pretrained model is fine-tuned with the new data. If the new data set is small and similar to the data set on which the pretrained model has been trained, then the weights of the first few layers or all the layers except the existing classification layer are frozen, and the classification layer is replaced by a new classification layer fitting to the new problem and is fine-tuned on the new data set. The first few layers extract more generic features like edges and curves and progressively along the network the features become more data-specific. So depending on the data set size and similarity of the new task with the task for which the pretrained network was trained, one must choose the layers for fine-tuning. It is also important to choose a much smaller learning rate for fine-tuning since the pretrained weights are better than random initialization and should not be distorted heavily.

2.4 LSTM

MLP and CNNs cannot directly address the problem of information propagation through time. Several applications such as gesture sensing and tracking require the neural network to keep a history of the past events to make a decision in the future. Recurrent neural networks (RNNs) address this issue by having self-loop structures, allowing information to persist over time. Using these self-loops, RNNs are able to connect previous information to the present task and make a decision by relying on past events. One of the biggest challenges toward wide adoption of RNNs in the 1990s was the problem of the vanishing gradient. If the weights are less than 1, then further multiplication with another parameter less than 1, would lead to a very small number after a few multiplications. Thus, the gradient flows over time through RNNs would easily become zero, which means no further propagation of information. Thus, RNNs were unable to retain information or learn information from the distant past/time.

RNNs can be described as follows: given a temporal input sequence $\mathbf{x}(k) = (x_1(k), x_2(k), x_3(k))$, a RNN maps it to a sequence of hidden values $\mathbf{h}(k) = (h_1(k), \cdots, h_T(k))$ and outputs a sequence of activations $\mathbf{a}(k + 1) = (a_1(k + 1), \cdots, a_T(k + 1))$ by iterating the following recursive equation:

$$\mathbf{h}(k) = \sigma(W_{hx}\mathbf{x}(k) + \mathbf{h}(k - 1)W_{hh} + \mathbf{b}_h), \qquad (2.27)$$

where σ is the nonlinear activation function, $\mathbf{b}_h$ is the hidden bias vector, and W terms denote weight matrices, W_{hx} being the input-hidden weight matrix and W_{hh} the hidden-hidden weight matrix.

The activation for these recurrent units is defined by:

$$\mathbf{a}(k + 1) = h(k)W_{ha} + \mathbf{b}_a, \qquad (2.28)$$

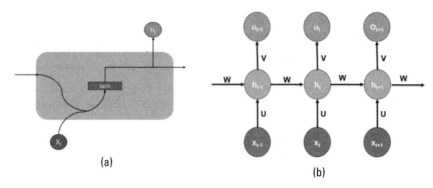

(a)

(b)

Figure 2.23   Illustration of operations in (a) basic RNN, and (b) RNN backpropagation through time.

where W_{ha} denotes the hidden-activation weight matrix and the $\mathbf{b}_a$ terms denote the activation bias vector. Figure 2.23 presents the operations in a basic RNN and backpropagation rolled through time.

The problem of vanishing or exploding gradient of RNNs is solved through LSTM [25] or gated recurrent unit (GRU) [26]. LSTMs extend RNN with memory cells using the concept of gating: a mechanism based on component-wise multiplication of the input that defines the behavior of each individual memory cell. The LSTM updates its cell state according to the activation of the gates. The input provided to a LSTM is fed into different gates that control which operation is performed on the cell memory: write (input gate), read (output gate) or reset (forget gate). These gates act on the signals they receive, and block or pass information based on its strength and importance, by virtue of their own learned filter weights. These weights are learned during the learning process, which means the weights of the cells learn when to allow data to enter, retain, or be deleted.

The vectorial representation (vectors denoting all units in a layer) of the update of an LSTM layer is as follows:

$$\begin{cases} \mathbf{i}(k) & = \sigma_i(W_{ai}\mathbf{a}(k) + W_{hi}\mathbf{h}(k-1) + W_{ci}\mathbf{c}(k-1) + \mathbf{b}_i) \\ \mathbf{f}(k) & = \sigma_f(W_{af}\mathbf{a}(k) + W_{hf}\mathbf{h}(k-1 + W_{cf}\mathbf{c}(k-1) + \mathbf{b}_f) \\ \mathbf{c}(k) & = \mathbf{f}(k)\mathbf{c}(k-1) + \mathbf{i}(k)\sigma_c(W_{ac}\mathbf{a}(k) + W_{hc}\mathbf{h}(k-1) + \mathbf{b}_c) \quad (2.29) \\ \mathbf{o}(k) & = \sigma_o(W_{ao}\mathbf{a}(k) + W_{ho}\mathbf{h}(k-1) + W_{co}\mathbf{c}(k) + \mathbf{b}_o) \\ \mathbf{h}(k) & = \mathbf{o}(k)\sigma_h(\mathbf{c}(k)) \end{cases}$$

where $\mathbf{i}$, $\mathbf{f}$, $\mathbf{o}$, and $\mathbf{c}$ are, respectively, the input gate, forget gate, output gate, and cell activation vectors, all of which are the same size as vector h defining the hidden value. The term σ represents nonlinear functions. The term $\{\mathbf{x}(1), \mathbf{x}(2), \cdots, \mathbf{x}(K)\}$ is the input to the memory cell layer at time k. W_{ai}, $W_{hi}, W_{ci}, W_{af}, W_{hf}, W_{cf}, W_{ac}, W_{hc}, W_{ao}, W_{ho}$, and W_{co} are weight matrices,

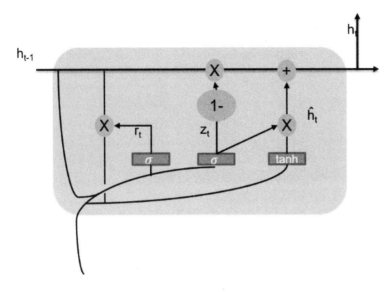

Figure 2.24 One LSTM unit.

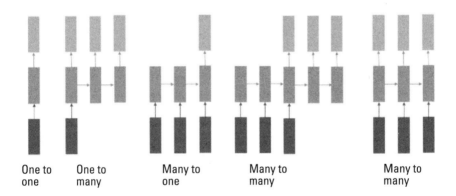

One to   One to   Many to   Many to   Many to
one   many   one   many   many

Figure 2.25 Various modes in which a LSTM unit be connected and used.

with subscripts representing from-to relationships b_i, b_f, b_c, and b_o are bias vectors. The modifications to a basic RNN in a single LSTM unit is presented in Figure 2.24.

Figure 2.25 presents the various configurations, such as one-to-one mapping, one-to-many, many-to-one, and many-to-many mapping, in which an LSTM unit can be used for various tasks [27]. GRUs are similar to LSTM with forget gate but lack output gate. The performance of GRUs on small data sets have been shown to be better than that of LSTM, but for a large-scale data set, LSTM performs much better.

2.5 Autoencoders

Autoencoders [28] work by transfering the input into a latent space representation with much less dimension (encoding) and reconstructing back the input as output from this representation (decoding). Thus, an autoencoder has two parts:

1. *Encoder*: The encoder basically compresses the input into latent space representation and can be mathematically represented as $h = f(x)$.
2. *Decoder*: The decoder tries to reconstruct back an output as close as possible to the input and can be mathematically represented as $r = g(h)$ where r is very close to x in ideal case.

So the entire autoencoder can be represented as $r = g(f(x))$. Figure 2.26 presents the autoencoder with the encoder and decoder structures. The latent space representation has many useful properties. One example is that it can act as a feature extractor if we limit the dimension of the latent space to a value smaller than the input, then we will have important features that exhibit favorable properties and also act as a compressor. However, one must be careful while using autoencoder for compression as it can only compress information efficiently for the type of data on which it has been trained.

There are different types of autoencoders:

1. *Undercomplete autoencoders*: These consist of a smaller number of neurons in the hidden layer compared to the input layer, so that they learn the most significant features in the data and are achieved by a loss function that minimizes the difference between input and output data. If we use linear activation and mean squared error to compute the difference between input and output, then we build a feature space that is quite similar to that of principal component analysis (PCA). However, if we use nonlinear activation functions then it becomes much more powerful. Figure 2.27 illustrates the

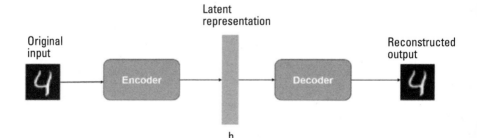

Figure 2.26   Basic autoencoder architecture.

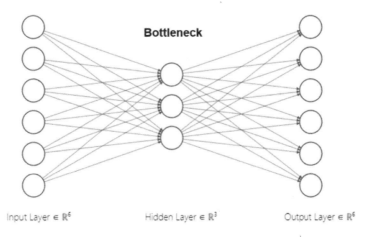

Figure 2.27 Undercomplete autoencoder architecture.

undercomplete autoencoder.

$$L = |x - \hat{x}|$$
$$L = |x - g(f(x))| \tag{2.30}$$

2. *Sparse autoencoders*: These [29] also perform the task of extracting features from the data. However, they have more neurons in the hidden layer than the input layer. To prevent overfitting, a sparsity constraint is added to the hidden layer and the sparsity penalty $\Omega(h)$, a value close to zero but not zero is added to the reconstruction loss function. Sparse autoencoders take only the highest activation values in the hidden layer and zero the rest. This is done to ensure that the model does not just learn to copy the data directly as the hidden layer has more neurons and rather each neuron learns to extract a feature from the data. Figure 2.28 presents the sparse autoencoder architecture.

$$L = |x - g(f(x))| + \Omega(h) \tag{2.31}$$

3. *Denoising autoencoders*: In these autoencoders, random noise is added to the input data. This can be easily done by randomly setting some of the input as zero and the remaining are fed as input to the network. The goal of the denoising autoencoder is to remove the corruption in the input data and reconstruct an output as close as possible to the original input data by minimizing the loss between the raw input and output instead of the input fed to the network. The network is a stochastic

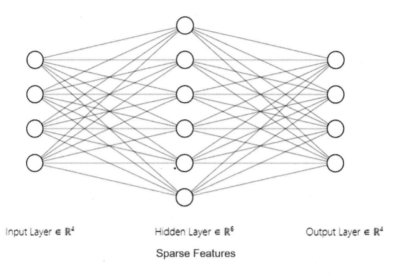

Input Layer ∈ **R**⁴   Hidden Layer ∈ **R**⁶   Output Layer ∈ **R**⁴

Sparse Features

Figure 2.28   Sparse autoencoder architecture.

autoencoder since we use a stochastic process to set some of the inputs as zero. Figure 2.29 illustrates the denoising autoencoder architecture.

4. *Contractive autoencoders*: These [30] are another regularization-based autoencoder that outperforms autoencoders based on weight decay regularization. The goal of such an autoencoder is to become insensitive to small variations in data. This is achieved by introducing a penalty term, the Frobenius norm of the Jacobian matrix, which is computed by calculating the squared sum of derivative of each hidden node with respect to input

$$L = |x - g(f(x))| + \lambda ||J_f(x)||_F^2$$

$$||J_f(x)||_F^2 = \sum_{ij} \left(\frac{\partial h_j(x)}{\partial x_i} \right)^2 \qquad (2.32)$$

2.6 Variational Autoencoder

The VAE [31] is a special type of autoencoder that is a generative model and estimates the probability distribution function (PDF) of the training data. However for a generative model, learning the PDF for each data attribute independently is not very helpful since the data attributes are mostly interdependent (e.g., images). In VAE training, to maximize $P(x)$ where $P(x) = \int P(x|z)P(z)dz$. $P(x)$ is the probability of a given input X, z is a latent vector in latent space, and $P(z)$ is the prior. $P(x|z)$ gives us information about whether X can be generated from that given latent vector z.

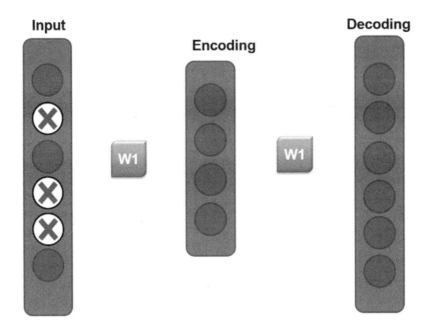

Figure 2.29 Illustration of denoising autoencoder architecture.

Since $P(x)$ is intractable, the Monte Carlo approximation can be used. However, for high-dimensional data, a lot of samples are required to have a good approximation. To model $Q(z|x)$ at the output of the encoder a multivariate Gaussian, a mean μ and covariance matrix Σ are estimated. In the case of VAE, a Kullback-Leibler (KL) divergence is used, which measures how different two distributions are. The entropy for a probability distribution is $H = -\sum_{i=1}^{N} p(x_i) \log(p(x_i))$.

The KL divergence is the expectation of the log difference between probability of data sampled from the approximating distribution to target distribution and thus is defined as

$$D_{\text{KL}}(p||q) = \sum_{i=1}^{N} p(x_i) \log\left(\frac{p(x_i)}{q(x_i)}\right) \tag{2.33}$$

KL divergence has the following properties:

1. KL divergence is 0 when both distributions are approximately the same;

$$D_{\text{KL}}(p||q) = 0 \quad \text{iff} \quad p \sim q$$

2. KL divergence is always positive for any two distributions;

$$D_{\text{KL}}(p||q) > 0 \quad \text{if} \quad p \neq q$$

3. To ensure $D_{KL}(p||q)$ is finite, the support of p needs to be contained in q, or by (2.33) $q(x) \to 0$ then $D_{KL}(p||q) \to \infty$

4. KL divergence is an asymmetric metric; that is,

$$D_{KL}(p||q) \neq D_{KL}(q||p)$$

thus $D_{KL}(p||q)$ is not a distance metric.

There are two variants of KL divergence:

- *Forward KL*: Say $q(x)$ is the normal distribution that is the target distribution and $p_\theta(x)$ is the approximating distribution where θ are parameterized as the weights of the VAE encoder. Then forward KL is defined as

$$\arg \min_\theta D_{KL}(q(x)||p_\theta(x)) \tag{2.34}$$

since the approximating distribution $p_\theta(x)$ must cover all regions of $q(x)$ thus $p_\theta(x)$ has a mean-seeking behavior as depicted in Figure 2.30(a).

- *Reverse KL*: The reverse KL is defined as

$$\arg \min_\theta D_{KL}(p_\theta(x)||q(x)) \tag{2.35}$$

in this case the approximating distribution is forced to be mode seeking since $q(x)$ must lie within a mode of $p(x)$, which is depicted in Figure 2.30(b). Thus reverse KL divergence is the prefered loss function for VAE; however, it must be noted that the results are still prone to error since even similar distribution shifted along x would have KL divergence tending to ∞ if the two distributions do not have any overlap.

In its original form, VAEs sample from a random node z, which is approximated by the parametric model $q(z|\mu, \Sigma)$ of the true posterior. In the decoder, since the encoded representation is not a sample but a distribution, and since sampling is not differentiable, backpropagation isn't possible and thus, encoder weights will not be updated. This is solved by the reparameterization trick in VAEs. The reparameterization trick in VAE transforms this random node z into a deterministic node by introducing a new parameter ϵ, as depicted in Figure 2.31, which is drawn from a standard normal distribution with mean 0 and standard deviation 1, which multiplies the square root of variance Σ output of the encoder.

Concretely,

$$\mu, \Sigma = Enc(x)$$

$$\epsilon \sim \mathcal{N}(0, 1)$$

$$z = \mu + \epsilon \Sigma \tag{2.36}$$

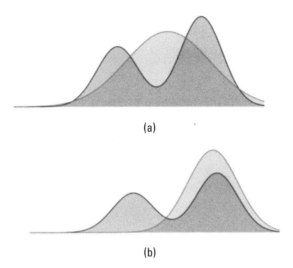

(a)

(b)

Figure 2.30   (a) Mean-seeking behaviour due to forward KL divergence, and (b) mode-seeking behaviour due to reverse KL divergence.

Thus, the reparameterized z allows for backpropagation to flow through the deterministic nodes. We can summarize the working of VAE with the following components:

- *Encoder*: Outputs latent vectors μ, Σ approximating the distribution $q(z|x)$;
- *Decoder*: Reconstruct input x_r from a latent vector sampled from $q(z|x)$;
- *Loss*: KL Loss + Reconstruction loss.

The reconstructed input data x_r is learned through the decoder part, which learns $x_r = p_\theta(x|z)$. The outcome of the trained model is that we can directly sample a latent vector z and decode it to generate new images. Figure 2.31 illustrates the variational autoencoder architecture.

2.7 Generative Adversarial Network

GANs, introduced by Ian J. Goodfellow in 2014 [32], are a breakthrough in the field of unsupervised learning using neural networks. The training of a neural network for generative modeling is done by using an adversarial process, which allows two neural networks to compete with each other to try to model the data distribution implicitly. The technique is one of the most promising due to its capability of modeling high-dimensional distributions and a less computationally expensive training process when compared with previous unsupervised learning methods like VAE, Boltzmann machines, and others.

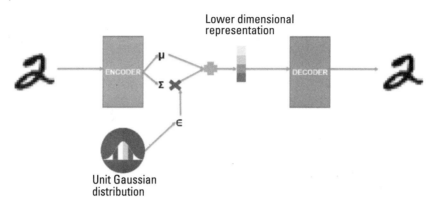

Figure 2.31   Variational autoencoder architecture depicting the reparameterization trick.

A GAN consists of a generator and discriminator. The task of the generator is to generate fake samples of the data, which are fed to the discriminator that tries to distinguish between the fake and real data. The training phase involves running both in competition; that is, the generator tries to fool the discriminator with fake generated image samples and the task of the discriminator is to identify the fake samples. Throughout the training process, both the networks get better and better in their respective tasks. Let's denote the inputs to the network as x with $x \sim \mathbb{P}_r$ where $\mathbb{P}_r$ represents the probability distribution of the real data. The generator in GAN tries to learn a distribution $\mathbb{P}_g$ that approximates the distribution of the real data. The generator takes as input a random variable sample (z) coming from a probability distribution $\mathbb{P}_z$. The trainable parameters are tuned over the training phase in order to minimize the distance or divergence between the distributions $\mathbb{P}_r$ and $\mathbb{P}_g$.

The objective of the discriminator is to maximize the expectation of the log likelihood of data drawn from real distribution; that is, $\max_D \mathbb{E}_{x \sim \mathbb{P}_r} \log(D(x))$, while minimizing the expectation of log likelihood of data generated from the generator, which samples from the random distribution $\mathbb{P}_z$; that is, $\min_D \mathbb{E}_{z \sim \mathbb{P}_z} \log(D(G(z)))$ or equivalently $\max_D \mathbb{E}_{z \sim \mathbb{P}_z} \log(1 - D(G(z)))$. It is easy to see that the latter maximizing function is a better optimization function. Thus the objective of the discriminator function is

$$\max_D \mathbb{E}_{x \sim \mathbb{P}_r} \log(D(x)) + \max_D \mathbb{E}_{z \sim \mathbb{P}_z} \log(1 - D(G(z))) \qquad (2.37)$$

On the other hand, the objective of the generator is to $\min_G \mathbb{E}_{z \sim \mathbb{P}_z} \log(1 - D(G(z)))$ so that the fake examples produced by the generator resemble the real data at the output of the discriminator. Thus, combining the two aspects and competing objective can be formulated as D and G are playing a minimax game

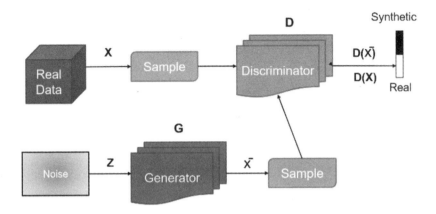

Figure 2.32 Vanilla GAN architecture outlining the principle of generator and discriminator.

with the combined loss function

$$\min_{G} \max_{D} \left[\mathbb{E}_{x \sim \mathbb{P}_r} \log \left(D(x) \right) + \mathbb{E}_{z \sim \mathbb{P}_z} \log \left(1 - D(G(z)) \right) \right] \qquad (2.38)$$

This is fine since $\mathbb{E}_{x \sim \mathbb{P}_r} \log \left(D(x) \right)$ is independent of the generator optimization. Figure 2.32 outlines the architecture of a GAN where a generator is competing against a discriminator to generate samples, which are not distinguishable from the real data at the discriminator. It can be shown that the generator is trying to minimize the Jensen-Shannon (JS) divergence between $\mathbb{P}_r$ and $\mathbb{P}_g$. JS divergence is bounded between 0 and 1 and is defined as

$$D_{\text{JS}}(p\|q) = \frac{1}{2} D_{\text{KL}} \left(p\|\frac{p+q}{2} \right) + \frac{1}{2} D_{\text{KL}} \left(q\|\frac{p+q}{2} \right) \qquad (2.39)$$

It is worth noting that unlike KL divergence used in VAEs, JS divergence is symmetric and in the case of two distributions being disjoint would result in a maximum value of $\log(2)$, irrespective of the two distributions when KL divergence would be ∞. From (2.39), it is easy to see that the minimum value of $D_{\text{JS}}(p\|q)$ is obtained when $p \sim q$ (i.e., in the case of a generator it is trying to achieve $\mathbb{P}_g \sim \mathbb{P}_r$ meaning, the generator generates samples that resemble the real data). The discriminator maximizes the loss by trying to approach $D(x)$ to 1 and $D(G(z))$ to 0, thus attaining the optimal value of $D^*(x) = \frac{1}{2}$, which is the Nash equilibrium.

Despite the remarkable contribution of GANs to the deep learning community as a whole, there are several problems with GAN, such as the training is not easy, it is slow, and it is unstable. Also another problem is that during training the generator may collapse to weights so that it is incapable of generating diverse data and is referred to as model collapse in the literature. Moreover, there is no way to calculate $\mathbb{P}_g(x)$ explicitly, so one mostly relies on visual comparison of generated data with real data for evaluation. Since the original work by Ian Goodfellow, a number of GAN-based architectures have been proposed suggesting several architectural changes, changes in loss function, procedural changes, and algorithmic inclusions to address several of these issues and we briefly describe some of the most popular architectures below.

1. *Conditional GAN (CGAN)*: In GAN, there is no means of controlling the labels or types of images the generator generates. This is overcome using CGAN [33], wherein labels y are added to latent vector z to introduce generations of distinctive data as per labels. Labels y are also fed as input to the discriminator in order to help the discriminator to distinguish between real and fake data. The label y is typically one-hot encoded and concatenated with random sample vector z and fed into the generator. Thus, the modified overall cost function is defined as

$$\min_{G} \max_{D} \mathbb{E}_{x \sim \mathbb{P}_r} \log \left(D(x|y) \right) + \mathbb{E}_{z \sim \mathbb{P}_z} \log \left(1 - D(G(z|y)) \right) \quad (2.40)$$

2. *Deep convolutional GAN (DCGAN)*: DCGANs [34] are one of the most successful implementations of GAN. They propose the following architectural changes—procedural and algorithmic changes—to stabilize GAN training:

   - Replace any pooling layers with strided convolutions (discriminator) and fractional-strided convolutions (generator);
   - Use batchnorm in both the generator and the discriminator;
   - Remove fully connected hidden layers for deeper architectures;
   - Use ReLU activation in the generator for all layers except for the output, which uses Tanh;
   - Use LeakyReLU activation in the discriminator for all layers.

   Further, DCGAN proposed a deduplication algorithm to remove near similar examples detected through a trained autoencoder, thus avoiding the problem of model collapse.

3. *Least square GANs (LSGAN)*: In the case of vanilla GAN, as mentioned when the real distribution $\mathbb{P}_r$ and the generated distribution $\mathbb{P}_g$ are disjoint, the JS divergence is log(2), which means the generator loss function does not learn anything as the derivative of constant tends to 0. LSGAN [35] proposes to use the least squares loss function to improve training during such issues. The loss function for generator G and discriminator D of a LSGAN are defined as

$$\min_{D} \frac{1}{2} \mathbb{E}_{x \sim \mathbb{P}_r} \left[\left(D(x|y) - b \right)^2 \right] + \frac{1}{2} \mathbb{E}_{z \sim \mathbb{P}_z} \left[\left(D(G(z) - a) \right)^2 \right]$$

$$\min_{G} \mathbb{E}_{z \sim \mathbb{P}_z} \left[\left(D(G(z) - c) \right)^2 \right] \quad (2.41)$$

where a, b are target discriminator labels for generated and real data, respectively, and c are the target generator labels for generator data. It can be shown different combinations of a, b, c can result in a cost function whose gradient is smooth, leading to the generator learning even when $\mathbb{P}_r$ and $\mathbb{P}_g$ are disjoint, unlike JS divergence, leading to $\mathbb{P}_g \to \mathbb{P}_r$.

4. *Wasserstein GAN (WGAN)*: In GANs, to achieve the Nash equilibrium the weights of generative and discriminative models are updated independently, resulting in high oscillation and training instability. When the real distribution $\mathbb{P}_r$ and the generated distribution $\mathbb{P}_g$ are disjoint, which typically is at the start of the training process, the JS divergence is constant; that is, $\log(2)$ resulting in no loss propagation and thus the generator does not learn. This is the drawback of JS divergence; another improvement proposed to overcome this issue is Wasserstein loss or earth-mover distance (EMD) in WGAN [36]. Intuitively, EMD measures the cost, measured as the amount of change times the changed distance required for transforming one probability distribution into the shape of another. The probability distribution is essentially a collection of earth, where the distribution measures the amount of earth at a given location. Thus, EMD can be represented as

$$\inf_{\gamma \sim \Pi(\mathbb{P}_r, \mathbb{P}_g)} \mathbb{E}_{(x,y)\sim\gamma} \left[||x - y|| \right] \qquad (2.42)$$

where $\Pi(\mathbb{P}_r, \mathbb{P}_g))$ are a set of all possible joint distribution between $\mathbb{P}_r$ and $\mathbb{P}_g$ and γ denotes all the possible transformations, thus the discriminator in this case finds the infinum inf (or lower bound) over all such transformations. However, it is easy to see that finding all transformations and then the lower bound is computationally infeasible. Thus, in WGAN the dual of the cost function is maximized such that

$$\frac{1}{K} \sup_{||f||_L \leq K} \mathbb{E}_{x\sim\mathbb{P}_r}[f(x)] - \mathbb{E}_{x\sim\mathbb{P}_g}[f(x)] \qquad (2.43)$$

where the function f is Lipschutz continuous and sup is the supremium or maximium bound. However, the constraint of the Lipschutz constraint is approximated by clipping the discriminator weights between $(-K, K)$ after every gradient update on the critic function f. Further, the WGAN attempts to train the critic f well before every generator update. In WGAN, the use of the Wasserstein distance instead of JS divergence as the loss function results in a smooth loss function even when the distributions are not overlapping.

5. *Laplacian pyramid GANs (LAPGAN)*: The Laplacian pyramid is a linear invertible image representation consisting of a set of band-pass images,

spaced an octave apart, plus a low-frequency residual [37]. In this type of GAN, multiple numbers of generators and discriminators are used for incorporating multiple levels of a Laplacian pyramid. The image is downsampled at each layer and then upsampled during a backward pass where it picks up some noise from the conditional GAN at these layers until it reaches its original size. This process results in the production of high-resolution images.

2.8 Robust Deep Learning

Deep learning has shown to perform remarkably well in the realm of computer vision tasks such as image classification, image segmentation, and detection. However, it has been shown in the literature that deep learning has an intriguing weakness such that when the images are perturbed by a small quantity of noise or adversarial attacks, which are imperceptible to human vision, the deep neural network can predict a completely different class or label with very high confidence. One school of thought is that the design of ReLu deep neural networks are linear at higher-dimension spaces and thus make them vulnerable to random perturbations. This aspect is crucial for sensors in general and also for radar sensors since the predictions rely on noisy measurements coming from the radar data. The search for robust deep learning architectures have lead to several approaches, such as autoencoder-based classification, distillation, Bayesian deep learning, which learns the classification and the associated uncertainity, and thus robust feature representation.

One approach to deal with adversarial attacks has been deep contractive networks, which use denoising autoencoders to reduce adversarial noise before classification. However, this approach has limited success. The other approach is distillation, where the idea is to transfer knowledge of a more complex network to a smaller network to handle adversarial attacks/noise during training. Alternately, Bayesian deep learning combines the Bayesian principles with deep learning and not only predicts the classification but also the uncertainty associated with the classification. Although Bayesian deep learning is able to achieve state-of-the-art results in computer vision tasks along with capturing the associated uncertainty, it achieves them at the cost of increased architecture computational requirements. The other alternative approach to achieve robust deep learning is to use robust feature images or vectors, which are not sensitive to noise perturbations.

There are a few metrics in the literature that evaluate the robustness lower bound of deep learning algorithms, namely Cross Lipschitz Extreme Value for Network Robustness (CLEVER) [38], CROWN [39], and CNN-Cert [40]. While CLEVER and CROWN work for MLP, although a CNN can be casted as a MLP architecture, CNN-Cert works directly on CNNs and are computationally faster to compute. Robustness is defined as the minimum adversarial perturbation of a given test point and a given trained neural network classifier that alters the predication. These metrics typically have their theoretical grounding in Lipschitz continuity of the classifier model. CROWN is a robustness certification algorithm to certify neural networks based on linear and quadratic bounding techniques on the activation functions and features adaptive activation bounds and can handle non-ReLU activations. The basic principle of finding these bounds are in an iterative process by bounding the first layer, and then bounds of successive layers are updated based on the bounds of the previous layers. These bounds are guaranteed to hold when the adversarial perturbation is bounded in Lp norm.

2.9 Problems

1. Compute the weights and bias for a single neuron that can implement the AND function. The activation function of the neuron is as $f(x) = 0$, for $x < 0$ else 1.
2. What is the purpose of introducing nonlinearities in neural networks?
3. Compare batch and stochastic gradient descent with respect to the requirement of shuffling training data.
4. Is initializing all the weights with the same value a good idea? Justify.
5. What are the advantages of using CNN over a MLP for an image classification task?
6. What are the hyperparameters for transfer learning?
7. Explain the advantages of LSTM over RNN.
8. Give example of a generative DL architecture and discriminative DL architecture.
9. How reliable are autoencoders for the task of data compression?
10. Explain the reparameterization trick in VAE. Explain the advantage of GAN loss over VAE loss.
11. Explain how the following model functions help in reducing model size; a. 1×1 convolution bottleneck layer, and b. squeeze and expand blocks.
12. What is mode collapse in GAN? How to identify a stable GAN training?
13. What are the improvements proposed by the Wasserstein GAN? How is the Lipschutz continuity implemented in a practical Wasserstein GAN?
14. What is bias-variance trade-off? What are the means of dealing with bias and variance in a neural network?
15. What is global receptive field in DCNN?
16. What is the difference between the approximation and machine/deep learning algorithms? Explain with an example.

References

[1]   LeCun, Y., Y. Bengio, and G. Hinton, "Deep Learning," *Nature*, Vol. 521, No. 7553, 2015. pp. 436–444.

[2]   Goodfellow, I., Y. Bengio, and A. Courville, *Deep Learning*, MIT Press, 2016.

[3]   Schmidhuber, J., "Deep Learning in Neural Networks: An Overview," *Neural Networks*, Vol. 61, 2015, pp. 85–117.

[4]   Rosenblatt, F., "The Perceptron: A Probabilistic Model for Information Storage and Organization in the Brain," *Psychological Review*, Vol. 65, 1958, pp. 386–408.

[5]   Widrow, B., and M. Hoff, "Adaptive Switching Circuits, *IRE WESCON Convention Record*, 1960, pp. 96–104.

[6]   Ruder, S., "An Overview of Gradient Descent Optimization Algorithms," ArXiv Preprint ArXiv:1609.04747, 2016.

[7]   Nair, V., and G. E. Hinton, "Rectified Linear Units Improve Restricted Boltzmann Machines," in *Proceedings of the 27th International Conference on Machine Learning (ICML)*, 2010, pp. 807–814.

[8]   Maas, A. L., A. Y. Hannun, and A. Ng, "Rectifier Nonlinearities Improve Neural Network Acoustic Models," *Proceedings of the 30th International Conference on Machine Learning (ICML)*, Vol. 30, 2013.

[9]   He, K., X. Zhang, S. Ren, and J. Sun, "Delving Deep into Rectifiers: Surpassing Human-Level Performance on Imagenet Classification," arXiv preprint arXiv:1502.01852, 2015.

[10]   Agatonovic-Kustrin, S., and R. Beresford, "Basic Concepts of Artificial Neural Network (ANN) Modeling and Its Application in Pharmaceutical Research," *Journal of Pharmaceutical and Biomedical Analysis*, Vol. 22, No. 5, 2000, pp. 717–727.

[11]   Sutskever, I., J. Martens, G. Dahl, and G. Hinton, "On the Importance of Initialization and Momentum in Deep Learning," in *Proceedings of the 30th International Conference on Machine Learning*, Vol. 28, No. 3, 2013, pp. 1139–1147.

[12]   Duchi, J. C., E. Hazan, and Y. Singer, "Adaptive Subgradient Methods for Online Learning and Stochastic Optimization, *J. Mach. Learn. Res.*, Vol. 12, 2011, pp. 2121–2159.

[13]   Tieleman T, and G. Hinton, "RMSprop Gradient Optimization," http://www. cs. toronto.edu/tijmen/csc321/slides/lecture?slides?lec6.pdf, 2014.

[14]   Zeiler, M. D., "Adadelta: An Adaptive Learning Rate Method," arXiv preprint arXiv:1212.5701, 2012.

[15]   Kingma, D.P., and J. Ba, "Adam: A Method for Stochastic Optimization," arXiv preprint arXiv:1412.6980, 2014.

[16]   LeCun, Y., and Y. Bengio, "Convolutional Networks for Images, Speech, and Time Series," in *The Handbook of Brain Theory and Neural Networks*, M. A. Arbib (ed.), MIT Press, 1995.

[17]   LeCun, Y., L. Bottou, Y. Bengio, and P. Haffner, "Gradient-Based Learning Applied to Document Recognition," *Proceedings of the IEEE*, Vol. 86, pp. 2278–2324, 1998, doi: 10.1109/5.726791.

[18] Iandola, F. N., S. Han, M. W. Moskewicz, K. Ashraf, W. J. Dally, and K. Keutzer, "SqueezeNet:AlexNet-Level Accuracy with 50x Fewer Parameters and 0.5 MB Model Size," arXiv preprint arXiv:1602.07360, 2016.

[19] Krizhevsky, A., I. Sutskever, and G. E. Hinton, "Imagenet Classification with Deep Convolutional Neural Networks," in *Advances in Neural Information Processing Systems*, 2012, pp. 1097–1105.

[20] Szegedy, C., W. Liu, Y. Jia, et al., "Going Deeper with Convolutions," *in Proceedings of the IEEE Conference on Computer Vision and Pattern Recognition (CVPR)*, 2015 pp. 1–9.

[21] Szegedy, C., V. Vanhoucke, S. Ioffe, J. Shlens, and Z. Wojna, "Rethinking the Inception Architecture for Computer Vision," in *Proceedings of the IEEE Conference on Computer Vision and Pattern Recognition (CVPR)*, 2016 pp. 2818–2826.

[22] He, K., X. Zhang, S. Ren, and J. Sun, "Deep Residual Learning for Image Recognition," *Proceedings of the IEEE Conference on Computer Vision and Pattern Recognition (CVPR)*, 2016, pp. 770–778.

[23] Tan, M., and Q. V. Le, 2019. "Efficientnet: Rethinking model Scaling for Convolutional Neural Networks," arXiv preprint arXiv:1905.11946.

[24] Pan, S.J., and Q. Yang, "A Survey on Transfer Learning," *IEEE Transactions on Knowledge and Data Engineering*, Vol. 22, No. 10, 2009, pp. 1345–1359.

[25] Hochreiter, S., and J. Schmidhuber, "Long Short-Term Memory," *Neural Computation*, Vol. 9, No. 8, 1997, pp. 1735–1780.

[26] Cho, K., B. van Merrienboer, C. Gulcehre, et al.," Learning Phrase Representations Using RNN Encoder-Decoder for Statistical Machine Translation," arXiv:1406.1078, 2014.

[27] Sherstinsky, A., "Fundamentals of Recurrent Neural Network (RNN) and Long Short-Term Memory(LSTM) Network," *Physica D: Nonlinear Phenomena*, 132306, 2020.

[28] Vincent, P., H. Larochelle, I. Lajoie, Y. Bengio, and P. A. Manzagol, "Stacked Denoising Autoencoders:Learning Useful Representations in a Deep Network with a Local Denoising Criterion," *Journal of Machine Learning Research*, Vol. 11, December 2010, pp. 3371–3408.

[29] Ng, A., "Sparse Autoencoder," *CS294A Lecture Notes*, Vol. 72, 2011, pp. 1–19.

[30] Rifai, S., P. Vincent, X. Muller, X. Glorot, and Y. Bengio, "Contractive Auto-Encoders: Explicit Invariance During Feature Extraction," *International Conference on Machine Learning (ICML) 2011*, Bellevue, WA.

[31] Doersch, C., Tutorial on Variational Autoencoders, arXiv preprint arXiv:1606.05908, 2016.

[32] Goodfellow, I., J. Pouget-Abadie, M. Mirza, et al., "Generative Adversarial Nets," in *Advances in Neural Information Processing Systems (NIPS)*, 2014, pp. 2672–2680.

[33] Mirza, M., and S. Osindero, "Conditional Generative Adversarial Nets," arXiv preprint arXiv:1411.1784, 2014.

[34] Radford, A., L. Metz, and S. Chintala, "Unsupervised Representation Learning with Deep Convolutional Generative Adversarial Networks," arXiv preprint arXiv:1511.06434, 2015.

[35] Mao, X., Q. Li, H. Xie, R. Y. Lau, Wang, Z. and S. P. Smolley, "Least Squares Generative Adversarial Networks," in *Proceedings of the IEEE International Conference on Computer Vision*, 2017, pp. 2794–2802.

[36]   Arjovsky, M., S. Chintala, and L. Bottou, "WassersteinGAN," arXiv preprint arXiv:1701.07875, 2017.

[37]   Denton, E. L., S. Chintala, and R. Fergus, "Deep Generative Image Models Using a Laplacian Pyramid of Adversarial Networks," *Advances in Neural Information Processing Systems (NIPS)*, 2015, pp. 1486–1494.

[38]   Weng, T. W., H. Zhang, P. Y. Chen, et al., "Evaluating the Robustness of Neural Networks: An Extreme Value Theory Approach," arXiv preprint arXiv:1801.10578, 2018.

[39]   Zhang, H., T. W. Weng, P. Y. Chen, C. J. Hsieh, and L. Daniel, "Efficient Neural Network Robustness Certification with General Activation Functions," *Advances in Neural Information Processing Systems (NIPS)*, 2018, pp. 4939–4948.

[40]   Boopathy, A., T. W. Weng, P. Y. Chen, S. Liu, and L. Daniel, "CNN-Cert: An Efficient Framework for Certifying Robustness of Convolutional Neural Networks," in *Proceedings of the AAAI Conference on Artificial Intelligence*, Vol. 33, 2019, pp. 3240–3247.

3

Gesture Sensing and Recognition

3.1 Introduction

Gesture sensing and recognition is the simplest, intuitive, and effective form of human–computer interface. Gesture sensing can replace interfaces such as touch and clicks needed for interacting with a device. Gesture recognition has applications in wearable and mobile devices, gesture-controlled smart TVs, smart home appliances, automotive infotainment systems, and augmented reality–virtual reality applications.

Traditionally, hand gesture recognition systems have been based on optical sensors and cameras [1]. Although optical sensors have a high resolution that enables tracking and recognition of the motions of the finger and wrist, they don't provide accurate depth estimates [2]. Camera-based hand gesture recognition can provide high accuracy, applying sophisticated computer vision techniques, such as hand segmentation, tracking, and classification [3]; however, such systems have limitations. They are limited by sufficient lighting condition requirements, suffer from self-occlusion issues, and can introduce privacy issues.

Recently, radar-based approaches for dynamic hand gesture recognition has attracted much attention from industry and academia [2,4–12]. Compared to vision-based gesture recognition systems, radar-based solutions are invariant to illumination conditions, hand visibility occlusions, and additionally provides privacy-preserving features together with capability to capture subtle hand gesture motions. Furthermore, the processing and classification pipeline for radar can be relatively less computationally expensive and thus facilitate embedded implementation. At the FMCW radar receiver, a hand gesture produces a superposition of reflections from different parts of the hand with different range and velocity changes over time, thus inducing a unique time-varying range-Doppler representation, which can help to detect and classify them reliably. In FMCW radar, a gesture recognition system typically consists of following major steps: gesture detection and preprocessing by creating the RDI while neglecting static targets and background environment. Feature extraction, which can be accomplished through handcrafted features such as Gabor transform or implicitly through a DCNN, and classification of the detected gesture from a library of trained gestures using conventional machine learning approaches such as random forest, and support vector machine, or deep learning approaches.

From an application point of view, gesture recognition using radars can be broadly classified into two categories, one using macro-gestures, which involve major hand/arm movement, and another using micro-gestures, which involve minute and subtle finger movements. Micro-gestures, applications are intended for detection and recognition of such gestures from ultrashort-range to a sensor with applications in car dashboards, watches, smartphones, or laptops. Macro-gesture applications are intended for detection and recognition of such gestures from a farther distance and find applications in smart projectors, smart TV systems, and smart lighting systems in rooms. Both categories have similar classification requirements but their respective sensing challenges are distinct and need to be addressed. In the case of micro-gestures, very minuscule finger movements, which typically cannot be detected through camera feeds, also need to be sensed and classified. Thus, macro-gestures recognition can rely only on gesture trajectory whereas micro-gestures also need to leverage micro-Doppler signature from the gestures. Radar signals are sensitive so they are capable of sensing such tiny motions and have granularity to distinguish such movements from others. On the other hand, in the case of macro-gestures there are interfering persons in the field of view that introduce interference to the sensor and are a source of missclassification. In the case of a micro-gesture sensing system the Doppler interference introduced by humans can be mitigated by reducing the transmit power or sensing only the nearest target. However, the same strategy doesn't work in the case of macro-gestures since the interfering human can be at the closer or the same distance to the valid gesturing human and thus different algorithmic and processing adaptations are required for enabling such applications using radars. Gesture recognition in a practical scenario is an open-set classification (i.e., the recognition system should classify correct known gestures while rejecting arbitrary unknown gestures during inference). However, conventional deep learning algorithms that generate the probability of a particular classification are not good enough to reject such arbitrary gestures or motions through softmax thresholding, thus requiring additional algorithmic changes. Figure 3.1 depicts a scenario between micro-gesture and macro-gesture applications.

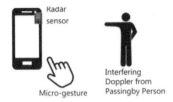

(a) Micro-Gestures Application

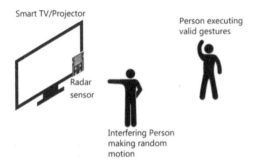

(b) Macro-Gestures Application

Figure 3.1 (a) Ultrashort-range micro-gesture applications, such as smart watches or smartphones, where typically the first detected target makes a valid intentional gesture and interference is at farther ranges. (b) Macro-gesture applications, such as smart TVs, smart lightning, and smart projectors, where the interfererer could be closer to the gesturing person and also the Doppler triggered by random motions [26].

3.1.1 Related Work

In [4], the authors use a 2.4-GHz Doppler radar to identify and distinguish several gestures using handcrafted features. The authors in [5] use a 24-GHz one transmit (Tx), two receive (Rx) FMCW radar to demonstrate gesture recognition by jointly calibrating the depth sensor. In [2], the authors propose a novel 3D deep convolutional neural network (3D-DCNN) for feature extraction, LSTM with connectionist temporal classification (CTC) loss function for the recognition of a gesture across time on camera, and depth data. In [6], the authors introduce a 60-GHz FMCW radar sensor with two Tx, four Rx channels and demonstrate reliable gesture recognition using handcrafted features and classification using a random forest classifier. In [7], the authors propose a novel signal processing pipeline to negate the effect of vibration facilitating gesture recognition using a random forest classifier in a car using 60-GHz FMCW radar. In [8], the authors propose a novel handcrafted feature using a sparsity-based approach for micro-Doppler extraction and subsequent recognition.

In [9], the authors proposed an end-to-end classification pipeline with 2D CNN with LSTM to implicitly extract features from range-Doppler images and recognize the gesture through the subsequent fully connected layers and LSTM layer and demonstrated the superiority of the approach using 60-GHz FMCW radar. The authors in [10] used a Doppler spectrogram from a Doppler radar and fed it into a DCNN to classify eight gestures with 85.6% accuracy. In [11], the authors extended the 2D CNN-LSTM approach to use long recurrent all-convolutional network (LRACN) to improve the accuracy, memory footprint, and computational complexity and improve inference time to facilitate customization into a real-time embedded platform. In [12], the authors propose to use 3D CNN-LSTM with CTC loss function using FMCW radar to improve the classification accuracy and support variable-length gestures by identifying the boundaries between gestures during inference, resulting in improved latency. In [13], the authors have proposed to use a phase range-Doppler map along with an amplitude range-Doppler map as input features and region proposal-based CNN for gesture sensing in a vehicle. Further, the authors demonstrate that the proposed approach is capable of recognizing overlapping and continuous gestures. In [14], the authors propose short-range gesture recognition using meta-learning, namely triplet loss function, which enables gesture recognition under alien environments and reject gestures that are not part of the training library.

The chapter is laid out as follows: Section 3.2 presents the topic of gesture sensing to detect the start and stop of a gesture for further classification/recognition processing. Section 3.3 presents the micro-gesture sets and the associated system parameters for sensing micro-gestures in a given setup. Sections 3.4, 3.5, and 3.6 presents three different approaches 2D All CNN-LSTM, 3D CNN and pseudo-CNN, and meta-learning, respectively, for gesture classification using the micro-gesture data set. Section 3.7 presents macro-gestures and the associated system parameters, and Section 3.8 presents an unguided attention-based 2D CNN-LSTM solution to classify macro-gestures. We conclude with future work and directions in Section 3.9.

3.2 Gesture Sensing/Detection

Figure 3.2 presents an overall gesture sensing and recognition pipeline that includes

1. *Radar signal preprocessing*: This step involves the data acquisition of ADC data across samples in a chirp and multiple chirps in a frame,

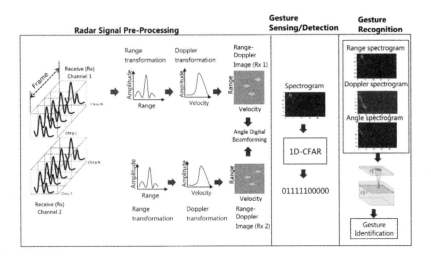

Figure 3.2 Overall gesture sensing and recognition pipeline including radar signal preprocessing, gesture sensing, or detection and gesture recognition.

followed by range transformation, Doppler transformation, and angle transformation. Based on the requirement, different features such as spectrograms or RDI are computed in this step.

2. *Gesture sensing/detection*: This step involves the detection of valid gestures for further recognition processing. This block detects either the start and end of a gesture with respect to time step/frame time or a bounding box around the valid gesture region in the spectrogram or RDI. Gesture sensing can be achieved through a classical signal processing approach through 1D CFAR algorithm or using the region-proposal network (RPN) to detect the bounding boxes.

3. *Gesture recognition*: This step involves feature image extraction and preparation and feeding into the neural network for classification. The feature image preparation includes extraction of relevant section of the preprocessed signal and preprocessing with zero-padding.

Figure 3.3 presents an example of a gesture sensing/detection pipeline. Post range-spectrogram removing responses from static target, the gesture detection could be achieved through either the 1D CFAR algorithm to determine the start and end timestamp of any gesture, or region proposal network (RPN), where a valid gesture region is detected through the proposal network. RPNs are typically used in faster-rcnn [15] to generate proposals for object detection. In a computer vision application, the candidate boxes are generated through the RPN and are classified to contain an object with probability along with outputting the bounding box of the target object. RPNs generate anchor boxes at each feature map pixel, also referred as anchor point, generated by the

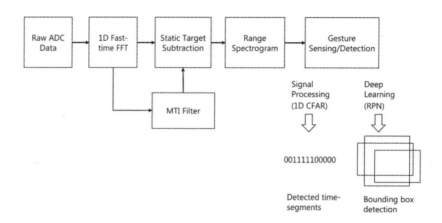

Figure 3.3 Gesture sensing and detection using either 1D CFAR algorithm or region-proposal network (RPN).

backbone network. At each anchor point, multiple anchor boxes of fixed and different dimensions are applied to generate the classification for that box and exact bounding box around the detected object. These exact bounding boxes are calculated by predicting offsets for x,y,w,h values, (x,y) being the center of the box, w and h are width and height, respectively. For a single target, it could so happen that multiple anchor boxes capture the same target. This is dealt with by using the intersection over union (IOU) metric (i.e., two bounding boxes with high overlap compared to total area are considered as that from the same object). Nonmaximal suppression (NMS) filter removes the overlapping bounding boxes for the same target, while retaining the one with the highest detection probability. In the case of radar, the same concept can be applied for estimating the bounding box of a valid gesture in the spectrogram feature image, as proposed in [13] and [16].

Alternately, instead of separate gesture sensing and detection processing through the 1D CFAR or RPN, this detection can be achieved through attention mechanism implicitly achieved within the gesture recognition neural network. Figure 3.4(a) presents an example RDI where gray cluster presents the target gesture region, such as fingers, dark pixels are noise, and whitish cluster represents interference. Figure 3.4(b) presents a soft-attentioned RDI, where the input RDI is multiplied with values between 0 and 1 through the sigmoid layer. The target cluster region is multiplied with values close to 1 and other noise and interference pixels with values close to 0 through soft-attention mechanism. Figure 3.4(c) presents a hard-attention RDI, where the target cluster pixels are multiplied with a value of 1 and nontarget pixels are multiplied with a value of 0 through a soft-attention mechanism. The attention mechanism is typically implicitly achieved

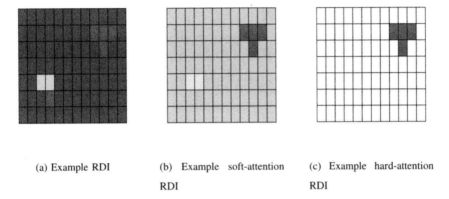

(a) Example RDI

(b) Example soft-attention RDI

(c) Example hard-attention RDI

Figure 3.4 (a) Example RDI, (b) soft-attentioned RDI, where target regions are multiplied with weights close to 1 and nontarget regions are multiplied to weights close to 0, and (c) hard-attentioned RDI, where target regions are multiplied with 1 and nontarget regions multiplied with 0.

within the classification network through parallel branch or fully-connected layer, thus avoiding the explicit gesture sensing/detection step. However, this comes at the expense of processing through the attention-enabled neural network at each time segment, whereas in earlier approaches the classification step can be skipped in the absence of a target.

3.3 Micro-Gestures

In this section, we discuss the applications of micro-gestures and how they are addressed through deep learning algorithms in practice.

3.3.1 System Parameters

Radar chipset BGT60TR24B, operating at 60 GHz with sweep bandwidth of 7 GHz is used, hence the theoretical range resolution of our system is $\delta r = c/2B = 2.14$ cm, where c is the speed of light. The fast-time window function results in some loss of range resolution. The number of transmit DAC samples used for generating the chirp is $NTS = 64$, hence the maximum detectable range is $R_{max} = (NTS/2) \cdot \delta r = 0.68$ m, since the sensor has only I channel. The ADC sampling frequency is set to 2 MHz, the maximum beat frequency is given by $\frac{2BR_{max}}{cT} = 495.83$ kHz.

The chirp time is set to 32 μs and the chirp repetition time is set to 64 μs and we use 16 consecutive chirps in a frame, thus the maximum velocity is given as $v_{max} = 39.0625$ m/s and the minimum velocity is $\delta v = 2.44$ m/s. The frame time is set to 20 ms, and 100 consecutive frames (i.e., 2 s of data is recorded once a gesture is detected). We use one Tx antenna and four Rx antennas for reception, and the 3-dB azimuth and elevation field of view (FoV) are both 70°.

3.3.2 Micro-Gesture Set

After a literature review on the use of different hand gestures in human-computer interaction, we have defined a gesture set for training purposes. All the chosen six gestures involve the movement of fingers or minimum muscle movement for hand displacement and are thus dynamic and micro. Gestures like finger rub and rotation allow us to exploit the intrinsic property of the radar, which, unlike a camera, does not suffer from self-occlusion and cannot recognize very small motion. The six selected gestures, whose RDI over time are illustrated in Figure 3.5, are

1. Grab (moving a hand toward the sensor, perform a grab action, and move it away);
2. Finger rub (displacement of thumb placed on the index finger);
3. Finger waves (movement of all the fingers above the sensor like playing a piano);
4. Circle (circular movement of a finger above the sensor);
5. Swipe (moving the hand horizontally from right to left);
6. Top-down (movement of the palm toward the sensor and away like pushing a button).

Figure 3.6 presents the equivaent range and Doppler spectrograms of the different micro-gestures.

3.4 2D All CNN-LSTM

One of the solutions to address the micro-gesture application is to use video of RDI from all the receive channels. In this case four receive channels are used as input images and use the time-distributed 2D all-CNN to act as a feature extractor generating a compact representation of the input RDI in a latent space. Once a gesture is detected, RDIs of all four channels from a frame are fed into the 2D CNN to generate this latent features and is done for all subsequent frames. The latent feature from each frame is then passed to the LSTM, which does the temporal modeling and generates the classification of the input gesture through full connected layer with softmax. Figure 3.7 presents the proposed pipeline, wherein a video of RDIs are generated once a gesture is detected and then the 2D all-CNN is used for feature extraction followed by LSTM for temporal sequence modeling coupled together to build an end-to-end model. After *argmax*(), the final classification label is passed to the application for specific action. Such end-to-end networks are referred as 2D all CNN-LSTM or LRACN since they only use convolutional layers as CNN for feature extraction. This architecture is favorable for an embedded solution due to the compact size and due to the fact that the data from RDI in each frame is consumed/compressed through 2D all-CNN without requiring it to be saved in a memory.

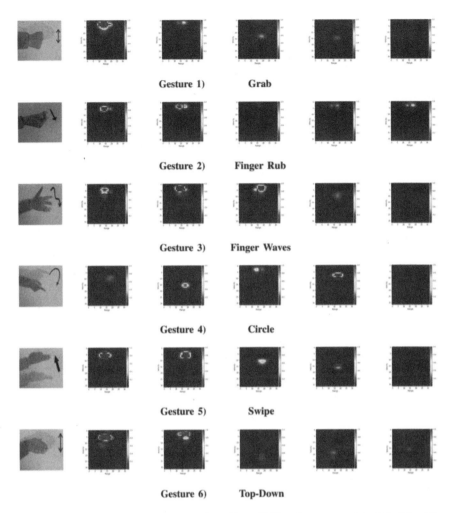

Gesture 1)   Grab

Gesture 2)   Finger Rub

Gesture 3)   Finger Waves

Gesture 4)   Circle

Gesture 5)   Swipe

Gesture 6)   Top-Down

Figure 3.5   Gesture set and its corresponding video of RDI at time stamps 0.1s, 0.3s, 0.9s, 1.3s, and 1.5s from detection of the gesture for Rx antenna one for (1) grab, (2) finger rub, (3) finger waves, (4) circle, (5) swipe, and (6) top-down gestures, respectively. In the x-axis, the negative velocity is represented by 0 to 15 bins and 16 to 31 bins represents positive velocity, whereas the y-axis represents increasing range bins from top to bottom.

3.4.1   Architecture and Learning

The LRACN model merges an intense hierarchical optical feature extractor (all-CNN) with a model that can utilize the knowledge of temporal dynamics for tasks involving sequential data. Figure 3.8 portrays the architecture of one of the proposed models. LRACN functions by passing each optical input x_t (a single range-Doppler image) through a feature transformation $v(.)$ done by all-CNN to generate a fixed-length vector representation $v(x_t)$. Generally, CNNs

Gesture   Range Spectrogram   Doppler Spectrogram

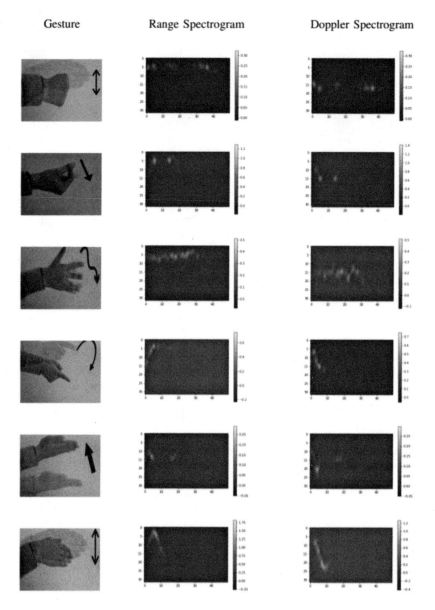

Figure 3.6   Range spectrograms and Doppler spectrograms of different micro-gestures for Rx antenna one for (1) grab, (2) finger rub, (3) finger waves, (4) circle, (5) swipe, and (6) top-down gestures, respectively.

are made of alternations of convolution layers and pooling layers with fully connected layers in the initial layers. In a DCNN, typically the most computations or trainable parameters are in the fully connected layers, thus 2D all-CNN benefits from the fact that it does not use fully connected layers. Further, many

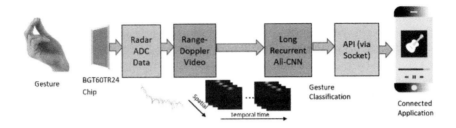

Figure 3.7 Overall gesture sensing pipeline [11].

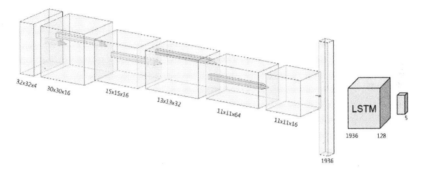

Figure 3.8 2D All CNN-LSTM Network architecture.

trainable parameters/weights mean that several training data sets are required to learn these parameters; otherwise, the network will overfit. Typically in the case of radar sensors available data is quite limited, thus the adaptation of all convolutional layers helps in achieving improved performance. CNNs with only sparsely connected layers (or convolutional layers) are called all-CNN. Since the number of parameters reduce due to the use of all-CNN, the model size also reduces.

The feature extraction is done by an all-CNN that comprises of three 2D convolutional layers, each with filter size of 3 × 3 and with ReLu activation followed by a dropout of 0.5, and a max-pooling layer with a stride of 2 and filter size 2 × 2 is used after the first convolution layer. Then, a convolution layer with a filter size of 1 × 1 with ReLu activation and dropout of 0.5 is present that extracts information across the channels. Batch normalization is done at every convolution layer. The temporal sequence modeling is done by using a LSTM layer that accepts a flattened output from the all-CNN of a sequence length of 100 frames. The outputs of $v(x_t)$ are later passed into a recurrent series learning module. In its most broad form, a recurrent model has parameters W and maps an input x_t and prior time step hidden state h_{t1} to an output z_t, and updates the hidden state to h_t. Therefore, inference must be run in series by calculating in order: $h_1 = f_w(x_1, h_0) = f_w(x_1, 0)$, then $h_2 = f_w(x_2, h_1)$, and so on, up to

h_t. To predict gesture class distribution, the output is flowed through a linear prediction layer, and ultimately the predicted class distribution is computed by the softmax operation. For T time steps, the above recurrence specifies that the last few predictions from the recurrent network has similar representation power to that of a T-layer deep network. To force the model to learn generic time-step-wise dynamics instead of dynamics conditioned on the frame or sequence index, the model's weights W are reused at every time step. In our use case, we have a sequential input and a static output. With sequential inputs and static outputs, we take a late-fusion approach to mix the per-time-step, foregoing $(y_1, y_2, ..., y_T)$ into a single prediction y for the full sequence. The LSTM layer uses tanh activation and hard sigmoid as recurrent activation. The LSTM layer is followed by a dense layer with softmax activation, which gives a probabilistic distribution over the five gesture classes for a given sequence.

Since the data is collected by a limited number of users, we added variance in the range-Doppler images for training. We propose a data augmentation technique that increases the data set size by creating synthetic images with variance to the original RDI to achieve generalization of the model. Initially for each gesture class and for each time step we form a mean range-Doppler image from all channels. Next, for each original record, we form a synthetic record by generating values for each pixel in time step t range-Doppler image by drawing values from a normal distribution with mean equal to the original Range-Doppler map at that time step and variance drawn from gesture class variation. This accounts for the time variations with which different individuals make the same gesture.

3.4.2 System Evaluation

For each gesture class, 150 sequences were recorded where each sequence contains 100 frames, making each gesture a few seconds long performed by 10 individuals with minimal prior instructions. A background range-Doppler image is saved when the system boots up with the set system configurations, and then the root-mean-square error (RMSE) between new range-Doppler maps and the background is calculated to detect the start of a gesture and then a sequence/video of range-Doppler image is flowed into the LRACN. The entire data set size is augmented by using a data augmentation technique as outlined in earlier Section 3.4.1. A training-validation data split of 80%–20% is performed on the data set for training and hyperparameter tuning. The testing data set of 600 sequences were collected from the other five individuals, who perform the same gestures. The kernel weights were initialized from a Glorot uniform distribution, and an Adam optimizer with a learning rate of 0.001 is used for cross-entropy minimization.

Table 3.1 presents the confusion matrix of the inference drawn on the testing data set from the LRACN model. In the table, gesture 1 refers to grab, gesture 2 referes to finger rub, gesture 3 refers to swipe, gesture 4 refers to up-down, and

Table 3.1

Confusion Matrix of Proposed LRACN Network [11]

		Predicted				
		Gesture 1	Gesture 2	Gesture 3	Gesture 4	Gesture 5
Actual	Gesture 1	85	0	10	3.33	1.67
	Gesture 2	0	100	0	0	0
	Gesture 3	0	3.33	96.67	0	0
	Gesture 4	0	1.67	0	98.33	0
	Gesture 5	0	8.3	0	0	91.7

Table 3.2

Analysis of Proposed Model in Terms of Model Size,
Accuracy, and Inference Time [11]

	Size	Accuracy	Inference Time
LRACN	13.16 MB	94.34 %	1s

gesture 5 refers to circle. As can be seen from the table, the finger rub gesture has 100% accuracy as in the RDI we get two blobs (one for the moving finger and the other one for the fist), which makes it very distinctive from all other gesture classes. We also observe that the circle gesture is sometimes confused with the finger rub, which is possible since both involve the movement of a finger and sometimes a circle gesture has two blobs with same RD intensity. Table 3.2 provides the model specification in terms of accuracy, model size, and inference time. All CNN-LSTM architecture greatly benefits from a very small model size (13.1 MB) and low number of floating operations (20 M) without model quantization and thus is easily scalable for edge neural processors for implementation on wearable and other small consumer devices. The 1×1 convolution layer for integrating the RDI data from across all virtual channels helps in increasing the accuracy by 2.34% and reduces the model size by 36 MB.

3.5 3D CNN and Pseudo-3D CNN

In 2D DCNN architectures, 2D convolutions are applied across 2D images to extract spatial features. However, in cases such as video or any sequence of 2D data, the temporal connections between the frames need to be modeled alongside the spatial information. In 3D CNN, the input is a 3D data cube formed by stacking multiple 2D data sequentially, resulting in a shape of $h \times w \times t$ where h is the height, w is the width, and t is the third dimension signifying the temporal axis.

3.5.1   3D CNN Architecture and Learning

A 3D CNN architecture majorly consists of the following components:

3D Convolutions: In 3D convolutions, a 3D kernel is convolved over the 3D data cube. In the ith layer, the value at position (x, y, z) in the jth feature images is given as

$$v_{ij}^{xyz} = \tanh\left(b_{ij} + \sum_m \sum_{p=0}^{P_i-1} \sum_{q=0}^{Q_i-1} \sum_{r=0}^{R_i-1} w_{ijm}^{pqr} v_{(i-1)m}^{(x+p)(y+q)(z+r)} \right) \tag{3.1}$$

where R_i is the size of the 3D kernel along the temporal dimension, and w_{ijm}^{pqr} is the (p, q, r)th value of the kernel connected to the mth feature image in the previous layer. These multiple filter kernels are required at each layer to extract diverse information.

3D Max-pooling: The 3D max-pooling layer allows to extract the most relevant spatial as well as temporal information, which helps in making the model invariant to small translations and also prevents overfitting to an extent.

Activation layer: The activation layer, also known as activation function, introduces nonlinear transformation on the input signal by learning whether a neuron will fire or not. ReLu is one of the most widely used activation functions. It facilitates faster backpropagation and doesn't activate all neurons and is therefore computationally very efficient.

Dropout: Dropout is a regularization approach that reduces interdependent learning among a set of neurons, thus preventing overfitting of training data.

Dense layer: The neurons in this layer have a complete connection to the high-level features extracted in the previous layers and their activation is computed by matrix multiplication and then a bias offset.

3.5.2   Pseudo-3D CNN Architecture and Learning

3D CNN kernels can be simply represented as $k \times k \times t$ where $k \times k$ is the spatial filter and t corresponds the temporal dimension. This can be decoupled as $k \times k \times 1$, which is similar to a 2D CNN in the spatial domain and $1 \times 1 \times t$ filters, which is similar to 1D convolution in the temporal domain.

Such a decoupled architecture can be perceived as a pseudo-3D CNN [17], which reduces the model size significantly and allows better scene understanding in spatial domain. The implementation of pseudo-3D CNN involves two major design issues, the first being whether the spatial and temporal filters should directly or indirectly influence each other. When the output of the spatial filter is fed as input to the temporal filter, then the filters have a direct influence on each other. In indirect influence, the spatial and temporal filters are decoupled in such a way that each have their own path in the network. The second issue is whether the filters should directly influence the final output.

The authors of the pseudo-3D DCNN propose the following architecture options:

1. In the first architecture, the output of the 2D spatial filters is directly fed as input to the 1D temporal convolution, which is directly connected to the final output.
2. The second architecture involves allowing both the filters to have separate paths in the network (indirect influence) and merging together to the final output.
3. The third architecture is comprised of a similar architecture as the first but additionally has a shortcut connection between the output of the spatial filter and final output that accumulates with the output of the temporal filter to the final output.

Figure 3.9 presents the three architectural option for implementing pseudo-3D DCNN.

The third architecture is well suited for our use case where there is a requirement of direct involvement of the spatial output to influence the final output as well as the requirement of temporal modeling.

3.6 Meta-Learning

In earlier sections, we explored several architecture, namely 2D-all CNN-LSTM, which used LSTM for temporal modeling and classification, and 3D CNN and pseudo-3D CNN, which feeds in video of RDIs where the temporal modelling is done through the 3D convolutional layers. However, such architectures alone with the conventional loss function, such as categorical cross entropy, still do not solve several of the practical challenges to be addressed for deployment of the

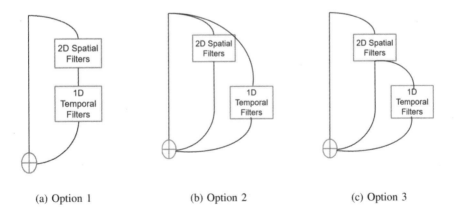

(a) Option 1   (b) Option 2   (c) Option 3

Figure 3.9   Pseudo-3D DCNN architecture with (a) option 1, (b) option 2, and (c) option 3.

product-ready solution. First, the gesture recognition system should be able to handle large interclass and intraclass differences of gestures. Interclass differences refer to the variation where the same gesture is performed by another person, such as someone making the same gesture slower and another faster. Intraclass differences refer to the variation arising when the same person performs the same gesture at a different instance and under different sensor orientation. Second, the gesture recognition system should be able to reject motions or unknown gestures. In a practical deployment, the system would be exposed to gestures or motions not trained for and the system is expected to infer these as random motions or gestures. Using conventional deep learning approaches, one approach of rejecting unknown gestures/motions is to apply a threshold on the output softmax probability. However, this is expected to work only for a very few subsets of cases [18], which is not acceptable for a product-ready solution. Third, the system should work under all alien or unknown background environments. The gestures sensing application is also expected to work under all background environments such as in a car's dashboard while perturbed with vibrations from the engine, in an unknown environment with interference from people walking nearby, background objects, and so on. For gesture sensing systems to work using conventional deep learning models under all plausible background environments, that would require a lot of training data for each class. However, modeling or collecting training gesture data in all variable environments and backgrounds, in most cases, is practically implausible. Therefore, conventional deep learning solutions would fail under such scenarios since little or no supervised information can be made available during training. Further, the system should be scalable and allow users to introduce gestures of their choice or convenience during the life cycle of its usage. However, conventional deep learning approaches first would require many training examples of the new gesture to be added. Second, once a model is trained for a library of gestures inclusion of a new gesture for inference would require retraining the whole network or fine-tuning of the last layers, which are computationally intensive and not conducive in an embedded platform without access to large compute resources.

To address the above challenges for a scalable and product-ready end-to-end gesture recognition system, meta-learning or learning-to-learn algorithms, specifically called few-shot learning [19,20] are used. Meta-learning using distance-metric-based models learns the relationship between data samples in the task space and thus are capable of adapting or generalizing to new tasks and new environments that have never been encountered during training time. To further learn appropriate feature representations and address the generalization capabilities, we exploit the triplet-loss-based embedding model [21]. The embedding model can simultaneously minimize the distance between similar gestures and maximize the distance between different gestures with a triplet loss function metric. Such a model and metrics have been successfully applied to a

2D image in the literature, and here is extended to 3D data, specifically video of radar RDIs. After the distance-metric-based model generates the gesture features embedding, the k-nearest neighbor (kNN) algorithm is used to recognize a known gesture even under an alien environment while rejecting unknown gestures using a simple thresholding technique to minimize false alarms. The embedding network learns a unique structure to naturally rank similarity between inputs. Once a network has been tuned, we can then capitalize on powerful discriminative features to generalize the predictive power of the network to not only new data from an existing class but to entirely new classes from unknown distributions through clustering (i.e., completely new gestures), without the requirement of retraining the embedding model. Furthermore, for training, embedding models require only a few example data compared to a large amount of data typically required for conventional deep learning approaches [22]. Using a 3D DCNN architecture in conjunction with triplet loss, we can solve practical issues and also achieve strong classification performance that exceeds those of other deep learning models with near state-of-the-art deep learning performance.

3.6.1 Architecture and Learning

The objective of the proposed 3D DCNN is to achieve an embedding $f(g)$ for a given gesture g (video of RDI) into a feature space R^d so that the squared Euclidean distance between same gestures is small and that of two different gesture is large, independent of the intrinsic properties of the radar. The model not only learns the class of g but also how different it is from other gestures g'. In terms of objective function this can be achieved through triplet loss distance-metric by training three 3D-DCNN sharing same weights feeding an anchor example, positive example (i.e., the same gesture), or negative example (i.e., different gestures). The triplet loss tries to form a margin between different gestures in the embedding space, thus allowing embeddings of the same gestures to exist in a close-knit cluster and distanced away from other gesture clusters.

Generally, 2D DCNNs are used to apply convolution on 2D feature images to extract features in the spatial dimension. However, in case of video of RDIs, temporal information that is present over a stream of input frames needs to be captured to make a sequential data classification. Therefore, we employ 3D convolutions in our network to capture spatial and temporal information. In normal 3D convolution, the input is a 3D cube where the third dimension denotes the temporal dimension containing sequence of frames and is formed by stacking 2D frames sequentially.

3.6.1.1 Triplet Loss Function

The embedding model embeds a RDI sequence g into a d-dimensional embedding space. The triplet loss used for the 3D data cube is inspired by

FaceNet [21] proposed in the context of 2D image face classifciation. Given the triplets (g^p, g^a, g^n) where p, a, n represents positive, anchor, and negative examples, respectively, we want to ensure that the anchor of a given gesture sequence is closer to all other sequences of the same gesture g^p than sequences of other gestures g^n and maintain a defined margin (α) between different gesture sequences. This requires the fulfillment of the following constraint:

$$||f(g_i^a) - f(g_i^p)||_2^2 + \alpha \leq ||f(g_i^a) - f(g_i^n)||_2^2 \tag{3.2}$$

$\forall f(g_i^a), f(g_i^p), f(g_i^n) \in T$ and the loss function can be defined as the following

$$\mathcal{L}_{triplet} = \sum_{i=1}^{N}[||f(g_i^a) - f(g_i^p)||_2^2 - ||f(g_i^a) - f(g_i^n)||_2^2 + \alpha]_+ \tag{3.3}$$

where $[x]_+$ represents $\max(0, x)$. Using all the possible triplets at every epoch would result in many triplets that satisfy the constraints after a few epochs. These easy triplets would not help in training and rather slow down the convergence as they will be passed through the network. Therefore, it is desirable to select hard triplets that would contribute to the training process and lead to faster convergence. The following subsection discusses the adapted triplet selection model. Figure 3.10 depicts the principle of triplet loss using 3D DCNN, wherein the objective function is to minimize the distance of the embeddings from anchor and positive examples and maximize the distance of the embeddings from anchor and negative examples.

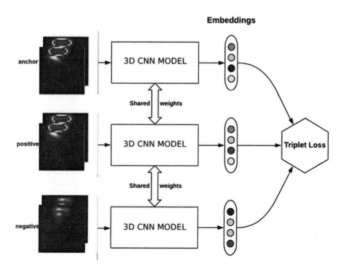

Figure 3.10   Principle of triplet loss based on the proposed 3D DCNN.

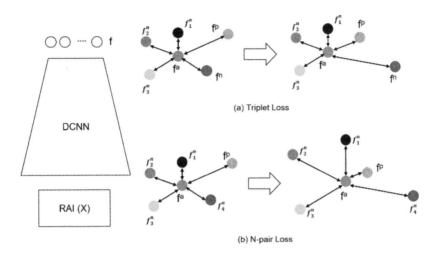

(a) Triplet Loss

(b) N-pair Loss

Figure 3.11 Principle of (a) triplet loss, and (b) N-pair loss.

Figure 3.11 illustrates the training representation for both triplet loss and multiclass N-pair loss. Multiclass N-pair loss function [23] is a generalization of the triplet loss, for which at each gradient descent update step utilizes all the other negative classes that the network should learn embedding vector. On the other hand, triplet loss utilizes just one negative example at each update step. The N-pair multiclass loss can be expressed as

$$\mathcal{L}_{N-\text{pair}} = \frac{1}{N} \sum_{i=1}^{N} \log(1 + \sum_{j \neq i} \exp(f_i^T f_j^+ - f_i^T f_i^+)) \qquad (3.4)$$

where f_i is the example embedding vector learned for ith class, f_i^+ is the embedding vector for the same class but for a different example, and f_j^+ is the embedding vector for a different jth class in the classification problem. Multiclass N-pair, by construction, generates a better embedding vector with better separation for each distinct class compared to triplet loss function at the expense of computationally expensive training.

3.6.1.2 Triplet Selection/Mining

Triplet selection or triplet mining is a critical aspect of training a DCNN using triplet loss, which, if not done correctly, the loss can get stuck in local minima after reducing drastically in the first few epochs. We adapt an offline triplet selection, where after every 100 epochs we save a checkpoint and select triplets that are semihard negatives. Semihard negatives are where the positive anchor distance is smaller than that of negative anchor but the negative anchor distance is close to the positive anchor distance and the negatives exist within the margin. We do not select the hard negatives as they can lead to early local minima and poor training.

3.6.1.3   Training Parameters

The weight initialization for 3D convolutional layers was performed by drawing samples from a normal distribution with zero-mean and standard deviation of 10^{-2}. The respective biases were initialized with samples drawn from a normal distribution also but with a mean of 0.5 instead of zero. The weights for the dense layers were initialized with a Xavier uniform initializer, which draws samples from a uniform distribution within [-limit, limit] where the limit is calculated by taking the square root of 6 divided by the total number of input and output units in the weight tensor.

An Adam [25] optimizer is used to compute an adaptive learning rate for each network weight over the learning process from estimates of first and second moments of the gradients. In configuration parameters, the learning rate (alpha) is set to 0.001 and the exponential decay rate for the first (beta1) and second (beta2) moment estimates are set to 0.9 and 0.999. The epsilon that counters divide by the zero problem is set to 1e-8.

3.6.2   System Evaluation

For each gesture, 100 consecutive RDIs are stacked together and fed to the network as an input shape of (100, 32, 32, 4). Triplet mining is performed and the (g^a, g^p, g^n) set is fed to the three identical 3D networks that share the same weights, as depicted in Figure 3.10.

Instead of linear activation, sigmoid activation is used at the last fully connected layer to prevent loss of information by yielding positive values. Different 3D architectures based on the trade-off of accuracy and model size have been explored. Our proposed solution, as depicted in Table 3.3, aims to run in small consumer electronics like mobile phones and wearable devices. Having deeper networks yielded slightly better results but increased the number of parameters and floating point operations exponentially. For our use case, the proposed model seems to be optimal with a model size of 7.96 MB and 3.1 MFLOPs. Further optimization using quantization, fusion, or weight-pruning will compress the model size. Once the embedding model is trained, a 1-NN classifier is trained to make gesture class prediction taking generated embedding of the gestures.

Typically, training deep neural networks requires a large amount of data; however, a deep neural network using triplet loss can be trained using much less data owing to the large number of possible triplet combinations. For each gesture class, 150 sequences were recorded where each sequence contains 100 frames making each gesture a few seconds long, performed by 10 individuals with minimal prior instructions. The data set was collected under different challenging

Table 3.3
3D Proposed DCNN Architecture Used for the Embedding Model

Layer (Type)	Output Shape	Parameters
conv3d_1 (Conv3D)	(None, 96,28,28,32)	16032
relu_1 (Activation)	(None, 96, 28, 28, 32)	0
max_pooling3d_1 (MaxPooling3D)	(None, 48, 14, 14, 32)	0
dropout_1 (Dropout)	(None, 48, 14, 14, 32)	0
conv3d_2 (Conv3D)	(None, 46,12,12,64)	55360
relu_2 (Activation)	(None, 46, 12, 12, 64)	0
dropout_2 (Dropout)	(None, 46, 12, 12, 64)	0
conv3d_3 (Conv3D)	(None, 44,10,10,64)	110656
relu_3 (Activation)	(None, 44, 10, 10, 64)	0
dropout_3 (Dropout)	(None, 44, 10, 10, 64)	0
max_pooling3d_2 (MaxPooling3D)	(None, 22, 5, 5, 64)	0
conv3d_4 (Conv3D)	(None, 18,1,1,64)	512064
relu_4 (Activation)	(None, 18, 1, 1, 64)	0
dropout_4 (Dropout)	(None, 18, 1, 1, 64)	0
conv3d_5 (Conv3D)	(None, 18,1,1,128)	8320
relu_5 (Activation)	(None, 18, 1, 1, 128)	0
dropout_5 (Dropout)	(None, 18, 1, 1, 128)	0
flatten_1 (Flatten)	(None, 2304)	0
dense_1 (Dense)	(None, 32)	73760
sigmoid_1 (Activation)	(None, 32)	0
Total Parameters		776,192

environments; in particular, gestures were recorded in a crowded background. Other environments included where the chip was hand-held and a gesture was performed with the other hand, as well as where the chip was placed in the car's infotainment dashboard while the engine was kept on and a copassenger made random movements. The training data set size was augmented by techniques mentioned earlier, resulting in a total training size of 1,800 (300 sequences for each class). A testing dataset of 420 sequences (70 sequences for each class) was recorded by five other individuals performing the same gestures under the same background environment as the training data set but 20% of the data was collected under different backgrounds and sensor orientation, which wasn't present in the training data set. A change in RDI energy is used to detect the start of a gesture and then a video of RDI is transferred into the 3D CNN to generate the embedding, which then goes into 1-NN classifier where it is either rejected as a false alarm or a gesture prediction is made. The end-to-end proposed radar-based gesture recognition system is depicted in Figure 3.12.

The overall end-to-end gesture sensing pipeline is evaluated by the following metrics:

1. *t-SNE representation [25]*: We use t-SNE for visualizing the generated embeddings in a 2-dimensional space, which gives a visually intuitive

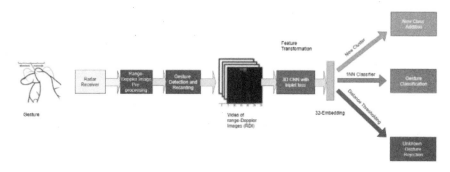

Figure 3.12   Overall gesture sensing pipeline depicting feature transformation through DCNN embedding model to enable gesture classification, rejection of unknown gestures, and new gesture class addition.

understanding about how efficiently the model clusters similar gestures together. t-SNE is a well suited nonlinear technique to visualize higher-dimensional data by performing dimensionality reduction. Initially, the algorithm computes the similarity probability of data points in input space and targeted lower-dimensional space. Next, the algorithm tries to minimize the conditional probabilities (or similarities) in both the spaces for a perfect lower-dimensional representation. The sum of Kullback-Leibeller divergence of all the data points is minimized using the gradient descent technique to measure the minimization of the sum of difference of conditional probability. One must note that after this process, the data points are no longer retrievable and hence the output of t-SNE can only be used for visualization and exploration purposes.

2. *Accuracy metrics*: Confusion matrices are computed to evaluate the recognition accuracy of our proposed end-to-end architecture. Further, the F1 score is computed by calculating the precision and recall to evaluate the capability of our proposed system to detect and reject false alarms.

3.6.2.1 Closed-Set Gesture Classification

The gesture classification capability of our proposed architecture is evaluated for known classes over the test data. The accuracy of the architecture is computed with yet another test set containing 180 sequences (30 each class) recorded with different background noise to evaluate the generalization capability of our embedding model. As seen in the confusion matrix depicted in Table 3.4, we see that our proposed model has high accuracy over all the classes and yields an overall accuracy of 94.5%. Furthermore, an interesting takeaway from the following result is how efficiently and accurately it can differentiate between very similar gestures like grab (gesture d) and top-down (gesture f). The result

Table 3.4

Confusion Matrix of the Test Set Containing
600 Examples from All Classes

		Predicted Class					
		a	b	c	d	e	f
Actual Class	a	90	0	0	0	10	0
	b	0	100	0	0	0	0
	c	0	3	92	0	5	0
	d	0	0	0	92	8	0
	e	0	3	0	0	97	0
	f	0	0	0	1	3	96

The overall accuracy is 94.5%.

Table 3.5

Comparison of Different Approaches for Gesture Classification under
an Alien Environment

Approach	Description	Accuracy
3D CNN	Categorical cross entropy loss function	86.3%
2D CNN-LSTM	CTC loss function	88.1%
3D CNN - kNN	**Triplet loss embedding**	**94.5%**

also shows that the embeddings generated by the proposed model is invariant to background noise to a great extent.

Further, Table 3.5 presents the classification accuracy of the proposed approach in comparison to known 3D CNN and 2D CNN-LSTM approaches. Since the test data set contains different background noise not present in the training data set, the proposed approach outperforms both the 2D CNN-LSTM model and 3D CNN model without any significant increment in model size. This behavior is well expected and can be explained because despite having a very small training data set, a huge number of triplet combinations are generated that allow the model to learn similarity features. This also allows high generalization capability of the model and is an intrinsic property of the proposed approach.

3.6.2.2 New Class Addition

To evaluate the scalability of our proposed embedding model, the trained embedding model is used for generating a few embeddings of the new class (not present in the initial training set). The trained embedding model projects the new gesture class in a unique cluster in the embedding space. The 1-NN classifier is then updated with the new class embeddings. This process allows adding a new

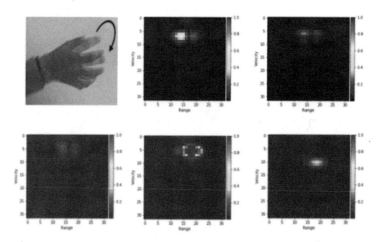

Figure 3.13   New rotation gesture and its corresponding video of RDI at time stamps 0.1s, 0.3s, 0.9s, 1.3s, and 1.5s from detection of the new custom rotation gesture for receive antenna 1.

class with a very few number of samples (in our case we used 15 samples) without the requirement to retrain the original deep learning model, which makes it a fast and computationally inexpensive approach toward expanding to include new gestures.

The existing test data used in the previous experiment along with 100 new rotation gestures (moving the hand like grabbing and rotating a knob, as shown in Figure 3.13) records were added, of which again 30 of them were recorded with unseen background not present in the training data set. Table 3.6 presents the confusion matrix generated for the (6+1) classes that show high accuracy for all gesture classes like in the previous experiment, including the new gesture. We achieve an overall accuracy of 94.57% and it is interesting to see how accurately the proposed model can recognize the new gesture rotation (gesture g), which was absent in the training phase of the embedding model without confusing it with a similar gesture such as finger waves (gesture c) present in the training set.

We use t-SNE to visualize the embeddings, in Figure 3.14 generated by our proposed model for a small subset of test data containing randomly chosen 10 example gestures for each gesture class including the new gesture class (rotation gesture). As we can see in the figure, each gesture class forms its own cluster as expected. However, the rotation gesture class and the finger-wave gesture class clusters are very close to each other with an outlier in each, since both the gestures are very similar with the involvement of all the five fingers. These observations confirm the unique embedding capability of our proposed model and the class expandable feature of the proposed architecture.

Table 3.6

Confusion Matrix of the Test Set Containing
700 Examples from All Classes When a New
Gesture is Added During Inference

		Predicted Class						
		a	b	c	d	e	f	g
	a	90	0	0	0	10	0	0
	b	0	97	0	0	0	0	3
Actual Class	c	0	3	92	0	5	0	0
	d	0	0	0	92	8	0	0
	e	0	3	0	0	97	0	0
	f	0	0	0	1	3	96	0
	g	0	0	0	2	0	0	98

The overall accuracy is 94.57%.

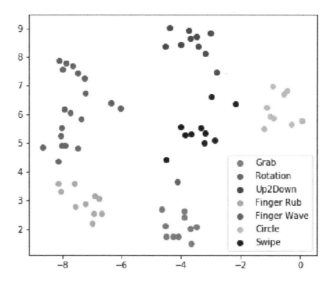

Figure 3.14   t-SNE representation of gestures embedding using the proposed 3D-DCNN using triplet loss.

3.6.2.3 Open-Set Gesture Classification

To detect unknown or invalid gestures/motions, the normalized l_2 distance (d) between the embedding of the input gesture sequence and the nearest neighbor embedding is computed and if the value is more than a defined threshold (d_{max}), it is tagged as a false alarm and thus 1-NN classification is not performed.

Table 3.7
Confusion Matrix of the Test
Set Containing 200 Examples
from All Classes

		Predicted	
		Invalid	Valid
Actual	Invalid	94	6
	Valid	7	93

The F1 score is 0.935.

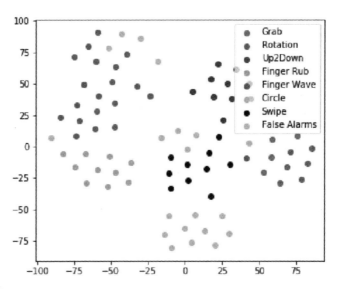

Figure 3.15 t-SNE representation of gestures embedding with invalid gesture using the proposed 3D DCNN.

The value of d_{max} can be tweaked according to the requirements, where setting a very low threshold would result in very strict acceptance of performed gestures and very high threshold would result in a very relaxed acceptance of the gestures. A relatively low threshold value was selected, which is optimal in our experiment and yields a precision of 0.93, recall of 0.94, and an F1 score of 0.935 computed from Table 3.7.

We also visualize the embeddings, in Figure 3.15, of the same subset along with 10 randomly chosen gesture data of invalid gestures. As expected, none of the invalid gestures (false alarms) are present inside any of the valid gesture clusters, thus demonstrating the invalid gesture or motion rejection capability of the system.

3.7 Macro-Gestures

3.7.1 System Parameters

The range and micro-Doppler information of the target is estimated by taking 1D FFT along the fast time, followed by 1D FFT along the slow time. The angle information is estimated by digital beamforming by multiplying the recieved range-Doppler data with phase shift sweeped across the field of view; that is,

$$RDI(\hat{\theta}) = \sum_{j=1}^{N_v} RDI^j \exp\left(-j\frac{2\pi d^j \sin(\hat{\theta})}{\lambda}\right)$$

$$\forall -\frac{\theta_{HPBW}}{2} < \hat{\theta} < \frac{\theta_{HPBW}}{2} \qquad (3.5)$$

where $\hat{\theta}$ is the estimated angle sweeped across the field of view; that is, $-\theta_{HPBW}/2 < \hat{\theta} < \theta_{HPBW}/2$, where θ_{HPBW} is the half-power beamwidth, and RDI^j, d^j is the complex RDI and position vector respectively from the jth virtual channel across N_v virtual channels. This transforms the 2D RDI across each virtual channel into RDI across the angle space. The mean RDI and RAI are generated by weighted mean across angle and Doppler, respectively. The weights are calculated by estimated signal power along those dimensions and is referred as maximal ratio combining (MRC). Finally, the mean RDI and mean RAI are thresholded using a constant false alarm rate detection algorithm to suppress noise.

Table 3.8 presents the system parameters used for our proposed gesture recognition system. The gestures voxel intensity represents the reflected energy from the target's hand, shoulder, and so on, in range, Doppler, and angle. The time-variant RDI and RAI captures the trajectories of the blobs over range, Doppler, and angle represented by the gesture's pattern. For the macro-gesture recognition system this dynamics across range, Doppler, and angle is captured by creating the radar data cube with a dimension of 32 × 32 × 32 across 32 range bins, 16 positive/negative Doppler bins, and 32 angle bins. Figure 3.16 presents the preprocessing steps to generate the mean RDI and mean RAI used as seperate channels for the macro-gesture system.

3.7.2 Macro-Gesture Set

The following eight gestures define macro-gestures that can be used to control TV/lightening/projectors:

1. *Anticircle*: Circular arm movements in a counterclockwise direction;
2. *Circle*: Circular arm movements in a clockwise direction;
3. *Cross*: Cross gesture like a virtual X drawn in air;
4. *Left to right swipe*: Swipes left to right with hand;

Table 3.8

System Parameter Summary [26]

Symbol	Parameter	Value
N_S	Number of ADC samples	64
N_C	Number of chirps	30
T_c	Chirp time	32 μs
PRT	Pulse repetition time	300 μs
N_{TX}	Number of transmit antennas used	1
N_{RX}	Number of recieve antennas used	2
B	Total bandwidth used	500 MHz
Δ_r	Range resolution	0.6 m
R_{max}	Maximum unambiguous range	19.2 m
Δ_v	Velocity resolution	0.26 m/s
v_{max}	Maximum unambiguous velocity	4.16 m/s
T_{frame}	Frame time	50 ms
Θ_{HPBW}	Azimuth antenna field of view	70°
P_{Tx}	Transmit power	10 dBm

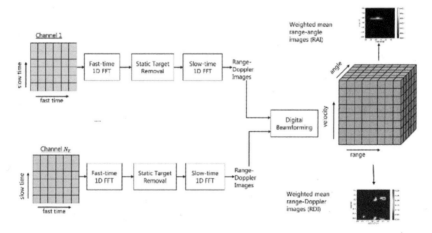

Figure 3.16   Preprocessing pipeline to generate mean range-Doppler-angle images and mean range-angle images [26].

    5. *Right to left swipe*: Swipes right to left with hand;
    6. *Push*: Push-button movement;
    7. *Pull*: Pull hand movement;
    8. *Wave*: Waving gesture.

Figure 3.17 presents the range-Doppler images and range-angle images due to the several gestures at different time steps from detection by a person at varying distance upto 6m away from the sensor. Using BGT60TR24B radar chipset operated with settings as specified in Table 3.8.

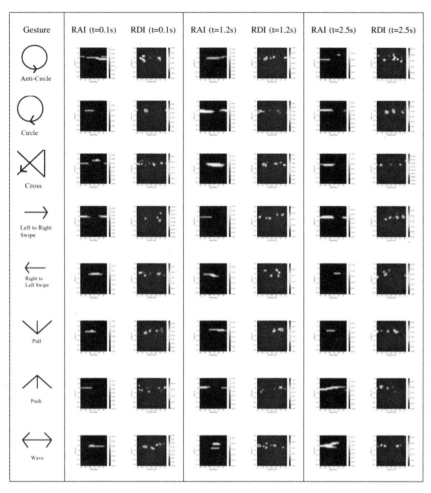

Figure 3.17 Gesture set and its corresponding video of RDIs and ADIs at 0.1s, 1.2s, and 2.5s from sensing the gesture for a counterclockwise circle, cross, left to right swipe, pull, and wave gestures, respectively [26].

3.8 Unguided Attention 2D DCNN-LSTM

3.8.1 Architecture and Learning

Time-distributed 2D-DCNN helps in encoding the spatio-Doppler information in the radar data cube over frames using learned local spatio-Doppler filters. The multiple spatio-Doppler filters learn useful constructs in several dimensions that facilitate attention of specific spatio-Doppler features and eventually classification of the different gestures. Figure 3.18 presents an 2D CNN-LSTM with a spatio-Doppler attention mechanism. The proposed model has a model size of 6 MB and 2.6 MFLOPs (without quantization or fusion or weight-pruning optimization, which are essential for embedded implementations).

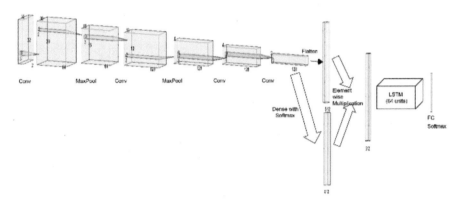

Figure 3.18   Proposed model architecture with spatio-Doppler self-attention [26].

Below are the descriptions of the unguided spatio-Doppler attention mechanism.

Unguided spatio-doppler attention $f(x)$ attention mechanism has been used widely in language translation problems in natural language processing (NLP) with relation to generating a context vector for an LSTM decoder. The attention mechanism has also been widely used in CNN for image captioning applications, and recently a self-attention mechanism was used in a generative adversial network for generating high-fidelity image synthesis.

In the case of computer vision problems such as these, there can be guided and unguided attention mechanism. The guided attention should provide a supervised label of the region where the neural network should focus and where it should reject in addition to having an overall loss function. In the case of unguided attention, the overall loss function dictates the attention map over which regions to focus and which to attenuate, and this is achieved unsupervised. In the latter case, through proper training and examples the attention mechanism would be able to learn the correct regions of importance and reject interference.

Attention is inspired by the way humans process an image. At any given time, instead of viewing an entire global image humans focus on some regions of the image to discern its interpretation. The problem with using the entire RDI for classification lies in the fact that the detected target's range-Doppler appears in a very small region compared to the entire image, thus introducing large noise for the classifier to classify the image reliably and thus lowering the detection accuracy.

In this case, the unguided spatio-Doppler attention mechanism is used for suppression of Doppler arising due to interference while focusing on a Doppler component due to the valid gesture to pass through to LSTM layer. Given the input feature map $x \in \mathbb{R}^{m \times n}$, the attention layer learns the feature weights $f(x)$, which is then elementwise multiplied with the feature map from the input convolutional layer. A seperate attention is learned along all the feature maps. The parallel attention layer through the fully connected layer acts a seperate neural network to emphasize and de-emphasize regions based on their spatio-Doppler properties.

In the case of spatio-Doppler attention mechanism, a simple dot-product based attention is applied. The global network is the classical convolutional neural network that processes the entire image and progressively shrinks the spatio-Doppler space. A parallel local branch is a mini-neural network based on the fully connected layer in conjunction with softmax to apply soft attention on the spatio-Doppler dimension applying different attention weights across all the feature channels. Since the convolution layer retains the relative position, thus attention applied through $f(x)$ followed by the softmax per feature map attenuates the Doppler due to interference and amplifies the Doppler due to valid gesture [26]. The attention mechanism learns to distinguish valid Doppler components from invalid ones effectively by the pattern of the Doppler. The effective attention per feature is given as

$$a_{i,j} = \frac{\exp(\alpha_{i,j})}{\sum_{i=1}^{m} \sum_{j=1}^{n} \exp(\alpha_{i,j})} \quad \alpha_{i,j} = f(x) \quad \forall i,j \tag{3.6}$$

The next layer recieves the feature maps where attention is applied. If the value of the per-channel attention weight at a certain spatio-Doppler position is close to 1, then these regions are focused compared to the ones whose values are close to 0. An alternative to continuous soft attention is to provide hard attention where the applied weights are binary (i.e., 0 or 1), which is similar to what is done in conventional signal processing. One of the main drawbacks of hard attention in comparison to soft attention is that the function is not differentiable and thus is not conducive for backpropagation. Following the self-attention-based 2D CNN, the temporal modeling of the embedding feature is performed by LSTM, which predicts the classification at the end of the fixed gesture length.

3.8.2 System Evaluation

For each gesture class, 200 sequences were recorded where each sequence contains 100 frames (i.e., 5s of recordings). We recorded the eight gestures by 10 individuals, wherein there were additionally none, one, or two other indiviuals in the field of view making random motions, such as working on computer, arbitrary hand movements, roaming around, or slight movements while sitting. The interfering persons were provided minimal supervision and were at different ranges and angles to the person making the valid gesture. The recordings were performed in a typical meeting room and drawing room. The training data set had 1,200 sequence and a testing data set of 400 sequences. A change in RDI

Table 3.9

Confusion Matrix of the Proposed Attention-Based 2D
CNN-LSTM with Overall Accuracy is 94.75%

		Predicted Class							
		a	b	c	d	e	f	g	h
	a	90	0	10	0	0	0	0	0
	b	0	96	0	1	2	0	1	0
Actual Class	c	2	1	97	0	0	0	0	0
	d	0	1	0	98	1	0	0	0
	e	0	0	0	1	99	0	0	0
	f	0	0	2	0	0	97	0	0
	g	0	3	2	0	2	0	93	0
	h	0	1	0	0	3	0	8	88

The gestures are as follows: (a) wave, (b) anticircle, (c)
circle, (d) cross, (e) left to right swipe, (f) pull, (g) push, and
(h) right to left swipe [26].

energy is used to detect the start of a gesture and then a video of RDI and RAI
are fed into attention-based 2D-DCNN with LSTM. We also use a dropout of
0.5 after each convolutional layer to prevent overfitting.

We trained 2D CNN-LSTM as proposed in Section 3.8.1 without
attention-layers using only range-Doppler images from both receive channels
on this far-distance gesture set including interfering Doppler and we achieved
an overall accuracy of 75.25% on the test data set. In contrast, the proposed
attention-based 2D CNN-LSTM achieved an accuracy of 94.75%. The confusion
matrix of the proposed solution is presented in Table 3.9. As expected, the
attention mechanism was capable of suppressing the Doppler component caused
by the interference person while allowing the Doppler component from valid
gesture that needed to be focused and passed through to the LSTM for temporal
modeling and classification. From the confusion matrix in Table 3.9, we observe
that the wave, circle, and anticircle gestures are often confused by the classifier,
most likely owing to the fact that the input features lack elevation information.
Furthermore, gestures such as push and pull are also confused at times, which is
mainly occurs because the proposed solution has poor range resolution operating
only at the 500-MHz ISM band.

The effect of using an attention layer in the network is illustrated in
Figure 3.19. For visualization, we use the output of attention layer for a given
input, which is in the shape (100,512) after dot multiplication and is reshaped
as (100,2,2,128). A mean is computed over the last axis resulting in a shape of
(100,2,2,1). Next, for a given time step we upscale the given 2×2 output 16

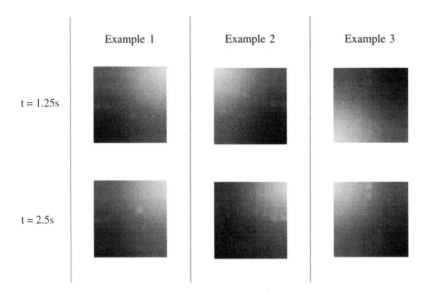

Figure 3.19 Attention map superimposed on the range-Doppler images for different examples [26].

times through a pyramid expand transformation to match the shape of input and superimpose it on the input image. As shown in the attention maps, for all the examples the attention map correctly focuses on the valid Doppler while rejecting the Doppler interference from the interferer. In the case of Example 2, at $t = 1.25$s the valid gesture is suppressed to some extent along with the interferer, however eventually the classification outputs are correct in these cases.

3.9 Future Work and Direction

In the case of micro-gesture sensing and classification using meta-learning, gesture classes are learned from the embedding vector through a 1-class nearest neighbor. Alternately, a fully connected layer with a softmax layer can be used to classify the different gesture classes. This solution is better for an embedded solution with a neural network accelerator, minimizing the data flow between the accelerator and micro-controller or digital signal processor (DSP). Furthermore, to improve the classification hyperplane a Silhouette clustering loss could be included in the loss function to generate better clusters in the embedding space.

In the case of macro-gesture sensing and classification, supervised spatio-temporal attention mechanism using a parallel branch on the feature image can be explored to achieve even better classification accuracy results. For supervised attention, the ground truth needs to be generated through manual annotation but provides better valid target gesture detection and classification.

3.10 Problems

1. What is the advantage of a RPN over an OS-CFAR detector for micro-gesture sensing/detection?
2. What are the possible disadvantages of using an all-CNN LSTM network?
3. What are the advantages and disadvantages of using spectrograms over videos of RDIs?
4. What are the advantages of meta-learning in the context of micro-gesture classification?
5. Why are hard negatives not selected in triplet mining?
6. Which feature image representation (video of RDI or spectograms) is more suitable for micro-gesture classification and why?
7. Is use of 3D CNN for RGB image classification a good idea? Justify.
8. What would happen if we used the ReLu activation function in the last layer of a triplet loss model?
9. State one advantage and one disadvantage of using a SVM instead of a 1-NN classifier in the meta-learning approach. How can we detect false alarms while using a SVM?
10. What is the advantage of pseudo-3D CNN compared to 3D CNN? What are the architecture options of pseudo-3D CNN?
11. How does dropout act as regularization? Is dropout a bagging or boosting technique?
12. Explain the importance of using attention mechanisms for macro-gesture classification.

References

[1] Mitra, S., and T. Acharya, "Gesture Recognition: A Survey," *IEEE Transactions on Systems, Man, and Cybernetics, Part C: Applications and Reviews*, Vol. 37, No. 3, 2007, pp. 311–324.

[2] Molchanov, P., X. Yang, S. Gupta, K. Kim, S. Tyree, and J. Kautz. "Online Detection and Classification of Dynamic Hand Gestures with Recurrent 3d Convolutional Neural Network," in *Proceedings of the IEEE Conference on Computer Vision and Pattern Recognition*, 2016, pp. 4207–4215.

[3] Rautaray,S. S., and A. Agrawal, "Vision Based Hand Gesture Recognition for Human Computer Interaction: A Survey," *Artif. Intell. Rev.*, Vol. 43, No. 1, 2015, pp. 154.

[4] Wan, Q., Y. Li, C. Li, and R. Pal., "Gesture Recognition for Smart Home Applications Using Portable Radar Sensors," in *Proc. 36th Annu. Int. Conf. IEEE Eng. Med. Biol. Soc.*, Aug. 2014, pp. 6414–6417.

[5] Molchanov, P., S. Gupta, K. Kim, and K. Pulli, "Short-Range FMCW Monopulse Radar for Handgesture Sensing," in *Proc. 2015 IEEE Radar Conf.*, May 2015, pp. 1491–1496.

[6] Lien, J., et. al., "Soli: Ubiquitous Gesture Sensing with mm-Wave Radar," *ACM Trans. Graphics*, Vol. 35, No. 4, 2016, Art. No. 142.

[7] Smith, K. A., C. Csech, D. Murdoch, and G. Shaker, "Gesture Recognition Using mm-Wave Sensor for Human-Car Interface," *IEEE Sensors Letters*, Vol. 2, No. 2, June 2018, pp. 1–4.

[8] Li, G., R. Zhang, M. Ritchie, and H. Griffiths, "Sparsity-Driven Micro-Doppler Feature Extraction for Dynamic Hand Gesture Recognition," *IEEE Transactions on Aerospace and Electronic Systems*, Vol. 54, No. 2, April 2018, pp. 655–665.

[9] Wang, S., J. Song, J. Lien, I. Poupyrev, and O. Hilliges, "Interacting with Soli: Exploring Fine-Grained Dynamic Gesture Recognition in the Radio-Frequency Spectrum," in *Proceedings of the 29th Annual Symposium on User Interface Software and Technology*, ACM, October 16, 2016, pp. 851–860.

[10] Kim, Y., and B. Toomajian, "Hand Gesture Recognition Using Micro-Doppler Signatures with Convolutional Neural Network," *IEEE Access*, Vol. 4, October 2016, pp. 7125–7130.

[11] Hazra, S., and A. Santra, "Robust Gesture Recognition Using Millimetric-Wave Radar System," *IEEE Sensors Letters*, Vol. 2, No. 4, December 2018.

[12] Zhang, Z., Z. Tian, and M. Zhou, "Latern: Dynamic Continuous Hand Gesture Recognition Using FMCW Radar Sensor," *IEEE Sensors Journal*, Vol. 18, No. 8, April 15, 2018, pp. 3278–3289.

[13] Sun, Y., T. Fei, S. Gao, and N. Pohl, "Automatic Radar-Based Gesture Detection and Classification via a Region-based Deep Convolutional Neural Network," *IEEE International Conference on Acoustics, Speech and Signal Processing (ICASSP)*, May 2019, pp. 4300–4304.

[14] Hazra, S., and A. Santra, "Short-Range Radar-Based Gesture Recognition System Using 3D CNN with Triplet Loss," *IEEE Access*, Vol. 7, 2019, pp. 125623–125633.

[15] Ren, S., K. He, R. Girshick, and J. Sun, "Faster R-Cnn: Towards Real-Time Object Detection with Region Proposal Networks," in *Advances in Neural Information Processing Systems*, 2015, pp. 91–99.

[16] Cai, X., J. Ma, W. Liu, H. Han, and L. Ma, "Efficient Convolutional Neural Network for FMCW Radar Based Hand Gesture Recognition," in *Adjunct Proceedings of the 2019 ACM International Joint Conference on Pervasive and Ubiquitous Computing and Proceedings of the 2019 ACM International Symposium on Wearable Computers*, September 2019, pp. 17–20.

[17] Qiu, Z., T. Yao, and T. Mei, "Learning Spatio-Temporal Representation with Pseudo-3D Residual Networks," *IEEE International Conference on Computer Vision (ICCV)*, October 2017.

[18] Bendale, A., and T. E. Boult, "Towards Open Set Deep Networks," *in Proceedings of the IEEE Conference on Computer Vision and Pattern Recognition (CVPR)*, 2016, pp. 1563–1572.

[19] https://lilianweng.github.io/lil-log/2018/11/30/meta-learning.html.

[20] Lake, B. M., R. Salakhutdinov, and J. B. Tenenbaum, "Human-Level Concept Learning Through Probabilistic Program Induction," *Science*, Vol. 350, No. 6266, 2015, pp. 1332–1338.

[21] Schroff, F., D. Kalenichenko, and J. Philbin, "Facenet: A Unified Embedding for Face Recognition and Clustering," in *Proceedings of the IEEE Conference on Computer Vision and Pattern Recognition (CVPR)*, pp. 815–823.

[22] Santra, A., I. Nasr, and J. Kim, "Reinventing Radar: The Power of 4D Sensing," *Microwave Journal*, Vol. 61, No. 12, 2018, pp. 26–38.

[23] Sohn, K., "Improved Deep Metric Learning with Multi-Class N-Pair Loss Objective," in *Advances in Neural Information Processing Systems*, 2016, pp. 1857–1865.

[24] Kingma, D. P., and J. Ba, "Adam: A Method for Stochastic Optimization,"arXiv preprint arXiv:1412.6980, December 22, 2014.

[25] Maaten, L. V. D., and G. Hinton, "Visualizing Data Using t-SNE," *Journal of Machine Learning Research*, Vol. 9, November 2008, pp. 2579–2605.

[26] Hazra, S., and A. Santra, "Radar Gesture Recognition System in Presence of Interference Using Self-Attention Neural Network," *2019 18th IEEE International Conference on Machine Learning and Applications (ICMLA)*, Boca Raton, FL, 2019, pp. 1409–1414.

4

Human Activity Recognition and Elderly-Fall Detection

4.1 Introduction

Human activity classification has many applications in smart homes, human-machine interfaces, and elderly fall-motion detection. Human activity recognition can enable better user intent sensing for smart homes and buildings. Elderly-fall motion detection and sensing can aid in elderly self-care in homes. Chen et al. developed mathematical and numerical models to analyze micro-Doppler effects of targets under translation, rotation, and vibration [1]. The worldwide elderly population aged over 65 years is growing and is projected to increase to 1 billion by 2030 [2]. Elderly fall has been identified as the leading cause of death among senior populations and such injuries can cause major impairment to mobility if not immediately attended to. Thus, technologies that can automatically identify elderly-fall motion and notify a neighboring medical facility will have profound social impact. Wearable sensors such as accelerators and radio frequency identifiers (RFIDs) are the most common fall detection systems. Alternate nonintrusive solutions include floor vibration sensors and video cameras that can monitor and detect elderly fall remotely. However, such solutions have their drawbacks—wearable sensors have to be worn all the time to be effective, are prone to damage, and cause false alarms, while the camera needs proper illumination to operate, is sensitive to occlusion in the environment, and leads to privacy concerns and thus is not a favored choice.

Radars, on the other hand, have emerged as a promising technology that can seamlessly and remotely sense human activities and identify fall motion. Additionally, radar brings in advantages of being insensitive to illumination conditions, maintaining privacy, and can be aesthetically concealed without affecting its operating performance. Different human motions generate distinguishable micro-Doppler effects in radar response that can be leveraged to recognize human activity, including fall motions. Micro-Doppler radar signatures have been shown to classify rigid targets, such as helicopters and aircraft [3], and wheeled and tracked vehicles [17]. In [5], micro-Doppler features have been shown to distinguish among humans, animals, and vehicles. These unique micro-Doppler signatures have been exploited for recognizing walking human and gait classification [6–9]. Several works have also focused on recognizing human activities such as walking, running, crawling, standing, and sitting [9–12]. A range spectrogram is the amplitude of the coherent summations of the complex time returns from target scatterers in each range cell, which represents the projection of the complex returned echoes from the target scattering centers onto the radar line-of-sight. The range spectrogram represents the target structure signatures, such as target size and scatterer distribution, and thus it has been used intensively in automatic target recognition applications [13–15]. Range spectrograms are in general sensitive to targets orientation angle and thus are not a reliable feature for robust target identification alone.

4.1.1 Related Work

Previously, proposed approaches rely on a conventional supervised learning paradigm based on handcrafted features. Feature engineering of radar returns has been well studied in the literature to classify human motion [6–12]. Recently, DCNN-based approaches have been shown to be effective for this problem, attaining equal or better performance than the conventional approach [12]. In [16], the authors explored the dynamic nature of a fall signal and used the Mel frequency cepstral coefficients, in conjunction with machine learning approaches, to differentiate radar echo between falls and nonfalls. In [17], the authors proposed to use six handcrafted features from a denoised radar Doppler spectrogram to classify seven different human activities, such as running, walking, walking with a stick, crawling, boxing forward, boxing in-place, and sitting still. In [18] multiple Doppler radars are exploited to raise the precision of fall detection by covering the field of view from multiple directions and to combat occlusions. In [19] wavelet transform based features followed by fuzzy SVM are proposed for elderly fall classification. In [20,21] the authors propose to use three handcrafted features—extreme frequency magnitude, extreme frequency ratio, and the time span of the event—for detection and classification of fall motion. DCNN can implicitly learn commendable features from training Doppler spectrogram without the need for explicit handcrafted features, and they can also significantly improve the classification accuracy. DCNN with different network architectures were proposed to recognize human activities [12,22]. In [23], the authors propose to use LSTM to model and characterize the temporal features in micro-Doppler signature for multitarget human gait classification. In [24], a deep neural network based on stacked autoencoders has been proposed to classify fall motion directly from Doppler spectrogram and has been shown to have superior performance compared to conventional machine learning approaches utilizing handcrafted features.

The chapter is laid out as follows: Section 4.2 presents the signal preprocessing for extraction of features and feature images, and Section 4.3 describes various feature images that can be used for classification of human activities using short-range radar. In Section 4.4, we describe and present a human activity data set collected using *BGT60TR24B* 60-GHz radar sensor and processed to generate feature images as mentioned in Sections 4.2 and 4.3. Activity classification using various DCNN architectures are presented in Section 4.5 along with their comparative study. In Section 4.6, we present a Bayesian classification approach for continuous human activity classification by integrating the classifier output to the tracker. The problem and challenges of fall-motion recognition is presented, as well as a solution using a deformable deep convolutional neural network (DDCNN) along with its loss function, and results are described in Section 4.7. We conclude with future work and directions in Section 4.8.

4.2 Preprocessing for Feature Image

When sensing a human, reflections from different parts of the human body, such as the torso, legs, feet, and hands, generate the radar echoes at the receiver. As features for human activity classification, either range spectrogram, Doppler spectrogram, or video of RDIs can be used. Although video of RDIs capture all the information, which is required for classification, they may not be the best solution for an embedded solution since the input dimensions are huge leading to a large-sized neural network for classification. Range spectrograms are not reliable features alone for activity classification; described later, Doppler spectrograms can suffer from several artifacts if the preprocessing to extract Doppler spectrograms is not done appropriately. In this section, we present a specialized signal processing pipeline to generate high-quality and clean Doppler spectrogram reducing the variations in Doppler spectrograms due to artifacts, thus enabling the classifier to scale in practical deployments. Figure 4.1 presents the overall digital signal processing pipeline that is used for extracting reliable, high-quality Doppler spectrograms with minimal artifacts for the classification step and is described in detail next.

4.2.1 Fast-Time FFT

The NTS number of transmit ADC samples from PN pulses in a frame are stacked into a PN × NTS matrix. The 1D FFT along fast time is computed for all PN pulses, which converts the fast-time data into range bins.

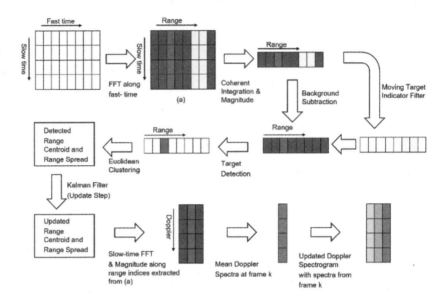

Figure 4.1   Preprocessing and generation of Doppler spectrogram [31].

4.2.2   Coherent Pulse Integration

After computing the FFT along the fast time for each chirp in a frame, pulse integration is applied to combine the data over all chirps to improve the SNR, faciliating subsequent target detection. We here apply coherent integration, which combines the phase and magnitude of the range FFT data coherently over all chirps of a frame. Coherent integration is based on the mean R_{ci} of the range FFT data r_i over PN chirps:

$$R_{ci} = \frac{1}{PN} \cdot \sum_{i=1}^{PN} r_i \tag{4.1}$$

$$r_i = a_i . e^{j\phi_i} \tag{4.2}$$

The amplitude and phase of the range FFT value for chirp i are denoted by a_i and ϕ_i respectively. However, phase misalignment leads to signal distortion in the coherent integration.

4.2.3   MTI Filtering

The idea of an MTI filter is to remove the static targets in the radar's field of view, since the target of interest—humans—will exhibit motion. At each frame, the coherent integrated range FFT spectrum is denoted as $R_{i,ci}$ and is passed through the running average as

$$R_{i,\text{ci}} = R_{i,\text{ci}} - S_{i-1} \tag{4.3}$$

$$S_i = \alpha \cdot R_{i,\text{ci}} + (1 - \alpha) \cdot S_{i-1} \tag{4.4}$$

where α is the forget factor set to 0.1. This essentially means that targets, which are static for more than several earlier frames, would be filtered out. This filtered range spectrum is then utilized for the subsequent target detection. The 1D range MTI with $\alpha = 1$ represents a highpass filter with a response of $1 - z^{-1}$ and thus having a highpass cutoff frequency given as the $f_c = 1/T_{\text{frame}}$. Thus, the frame time needs to be chosen accordingly, so that static targets are filtered out and the micro-Doppler components are retained.

4.2.4 Adaptive Detection Thresholding

The target detection is simply done through simple adaptive thresholding with a scaled value of the mean of the range FFT spectra calculated at every frame. Thus the threshold at frame i is set as $\beta \cdot \sum_{n=1}^{N_s} R_{i,\text{ci}}(n)$ where β is the scaling factor and is set to 3 in most cases.

4.2.5 Euclidean Clustering

Following target detection along the range axis, the target response from a single target is grouped together through the Euclidean clustering approach. The Euclidean clustering approach works by grouping together neighboring reflections from a single target based on the known length of the target, human in this case. If the distance between the selected detection and other remaining detections is less than the target length, then the detections are grouped together into the same cluster.

4.2.6 Kalman Filter

Target tracking maintains the state and identity of the detected human target over time accounting for detection errors (false negatives, false alarms), target occlusions, measurement noise, ghost targets, and the presence of other objects. The Kalman filter tracking is thus used for generating a Doppler spectrogram with minimal artifacts. The Kalman filter calculates the state vector x_k by recursively executing the predict step and the update step.

4.2.6.1 Process Model

The state variable is $x_k = \begin{bmatrix} r_k & \dot{r}_k & l_k & \dot{l}_k \end{bmatrix}$ where $r_k, \dot{r}_k, l_k, \dot{l}_k$ denotes the radial distance, radial velocity, range spread length, and change of range spread length at the kth frame. The measurement model is denoted by $z_k = \begin{bmatrix} r_k & l_k \end{bmatrix}$. The process model is given as

$$A = \begin{bmatrix} 1 & \delta t & 0 & 0 \\ 0 & 1 & 0 & 0 \\ 0 & 0 & 1 & \delta t \\ 0 & 0 & 0 & 1 \end{bmatrix} \tag{4.5}$$

where δt is the frame time or time duration between measurements.
The Kalman filter predict step involves the following steps:

$$\bar{x}_k = A x_{k-1}$$
$$\bar{P}_k = A P_{k-1} A^T + Q \tag{4.6}$$

4.2.6.2 Measurement Model
The observation or measurement model H can be described as

$$H = \begin{bmatrix} 1 & 0 & 0 & 0 \\ 0 & 0 & 1 & 0 \end{bmatrix} \tag{4.7}$$

The state vector x_k is updated as the following:

$$K_k = \bar{P}_k H^T (H \bar{P}_k H^T + R)^{-1}$$
$$x_k = \bar{x}_k + K_k (z_k - H \bar{x}_k)$$
$$P_k = (I - K_k H) \bar{P}_k \tag{4.8}$$

where K_k is the Kalman gain. The objective of the update step is to obtain the posteriori estimate of x_k with the linear combination of the a priori estimation and the new measurement z_k. In case of a missed detection, only the predict step is performed, omitting the update step.

4.2.6.3 Track Management
Once a target is detected over three consecutive frames a track is assigned to that target to avoid track assignments to ghost or spurious targets. The Doppler spectrogram is then recorded over a period of 3.15 s or 90 consecutive frames and then the fed into the DCNN. If the target is tracked beyond 3.15s another Doppler spectrogram is extracted. If the track is lost before 90 consecutive frames, the Doppler spectrogram is zero-padded before feeding into the DCNN.

4.3 Input Feature Images

The input feature images of the human activity classifier can be range spectrogram, Doppler spectrogram, or video of RDIs and are described here next.

4.3.1 Range Spectrogram

Range spectrogram captures the variation of the coherent range spectrum over time. The range spectrogram at frame time k is computed by applying 1D-FFT along the fast-time signal $s(l, k)$ as given as

$$R(m, k) = \left| \sum_{l=1}^{\text{NTS}} w(l)s(l, k)\exp\left(-\frac{j2\pi ml}{\text{NTS}} \right) \right| \tag{4.9}$$

where $w(l)$ is the window function, $s(l, k)$ is the fast-time data across for all PN chirps, and NTS is the number of transmit samples (i.e., data points along the fast time). The final range spectrum at frame k is computed by taking the coherent integration across all PN chirps in the frame followed by 1D MTI as explained in the Section 4.2.3. The target spectrum is characterized by the size and reflective properties of the scatterers comprising the target. Although, the range spectrogram is effective in capturing the geometric properties of the target, it suffers from the aspect angle variation and in some cases a small variation can lead to range bin migration and also due to a change in the differential path length, speckle or scintillation can be introduced due to the unresolved scatterers in a target. Owing to the superposition of different scattering components, the range FFT spectra present effects such as missed detections arising due to destructive interference, ghost target, arising due to interferences from scatterers, variation in a target's range-spread arising from different activities, and other anomalies. These effects are presented in the range spectrogram in Figure 4.2 due to human walking and human performing sitting up activities, respectively, and thus are not reliable features by themselves for practical target classification tasks.

4.3.2 Doppler Spectrogram

The Doppler spectrogram can be better estimated by extracting the averaged Doppler vector from the tracked range bins by the Kalman filter. The range bins, which are included in the range-spread length l_k and centroid r_k available from the state vector at each frame k, are used for extracting the average Doppler spectra for that frame. We denote $\mathcal{T}$ as the set that includes all these range bins. The Doppler spectrogram contains information about the instantaneous spectral content and the variation of the spectral content over time of a target. For human targets the waveform parameters can enable the Doppler spectrogram to capture both the macro-Doppler component due to the torso's movement as well as the micro-Doppler components generated due to hand, leg, and shoulder movements while executing a particular activity. Thus, the Doppler spectrogram can be particularly effective in classifying a target's activity or other properties present in the target's motion. The short-time Fourier transform (STFT) of the slow-time data from

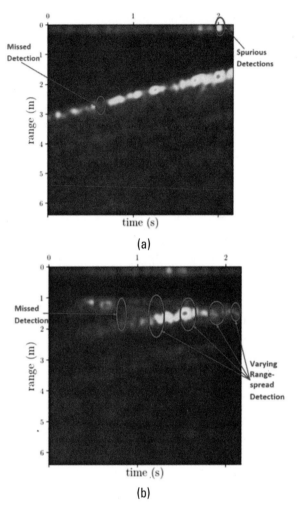

Figure 4.2   Range spectrogram of walking and sitting-up activities depicting missed detection, spurious detection, and varying range-spread detections [31].

the kth radar frame on the selected range bins can be expressed as

$$S(p; k)^m = \left| \sum_{n=1}^{PN} w(n)s(n; k)^m \exp\left(-\frac{j2\pi np}{N_{PN}}\right) \right|$$

$$S(p; k) = \sum_{m=1}^{M} S(p; k)^m \qquad (4.10)$$

where $w(n)$ is the window function, $s(n; k)^m$ is the slow-time data across PN chirps in the frame for mth range bin, and N_{PN} is the FFT points along slow time. $S(p; k)^m$ represents the Doppler spectrogram along the mth range bin, and summation of the Doppler spectrogram across $m \in \mathcal{T}$ range bins result in the desired Doppler spectrogram for the tracked target.

4.3.3 Video of RDI

The video of range-Doppler computed at frame time k can be expressed as

$$v_{\text{RDI}}(p, m; k) = \sum_{l=1}^{\text{NTS}} \left(\sum_{n=1}^{\text{PN}} s(n, l; k) w(n) e^{-j2\pi np/N_{\text{PN}}} \right) w(m) e^{-j2\pi ml/N_{\text{ADC}}}$$

(4.11)

where index l, n sweep along the fast-time and slow-time axes, respectively, while m, p sweep along the range and Doppler axes, respectively. $N_{\text{ADC}}, N_{\text{PN}}$ are the FFT size along the fast time and slow time, respectively. Figure 4.3 depicts the scenario where a video of RDI are captured along different time stamps as a human walks around.

4.4 Human Activity Data Set

We considered the following eight indoor human activities for the human activity classification task:

1. Sitting;
2. Walking;
3. Running;

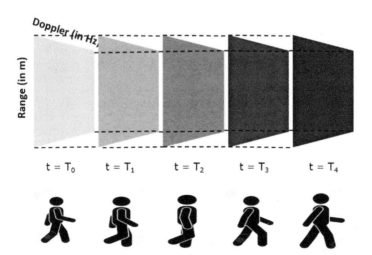

Figure 4.3 Video of RDI captured as a human walks.

4. Boxing while standing still;
5. Boxing while moving toward the sensor;
6. Waving;
7. Lying down;
8. Sitting up from a lying position.

The range spectrogram, Doppler spectrograms, and the corresponding video of RDIs were collected using *Infineon's BGT60TR24B*, which transmits at 60-GHz carrier frequency. The data set was collected with PN = 60 number of chirps with bandwidth $B = 1.5$ GHz, chirp time $T_c = 170$ μs, and number of ADC samples as NTS = 128. The chosen parameters result into a range resolution of 10 cm, a velocity resolution of 23.7 cm/s, and a maximum observable range and speed of 6.4 m and 7.1 m/s, respectively. The data was collected with transmit power of 10 dBm with one transmit and one receive antenna, and the gain of the reciever was set to 25 dB. From the detection of a target, 90 frames with a frame time of 35 ms (i.e., 3.15s of data windows are collected for activity recognition). Figure 4.4 displays the range spectrogram of the various everyday human activities. Figure 4.5 presents the corresponding Doppler spectrogram of the various human activities extracted using the preprocessing pipeline detailed in the earlier Section 4.3.2.

4.5 DCNN Activity Classification

4.5.1 Architecture and Learning

There are several DCNN architectures based on the input feature images that can be leveraged to classify human activities using radar sensors. Some of the typical

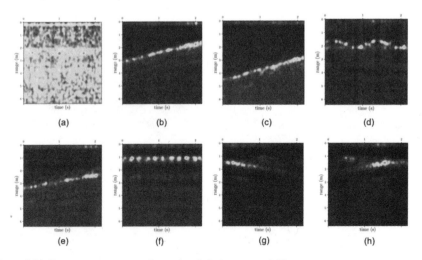

Figure 4.4   Range spectrogram of everday daily human activities.

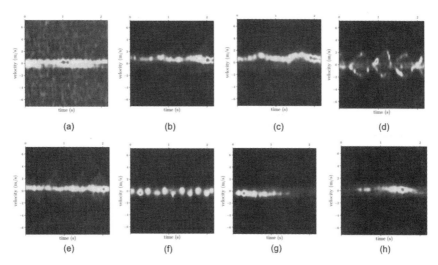

Figure 4.5 Doppler spectrogram of everday daily human activities: (a) sitting, (b) walking, (c) running, (d) boxing still, (e) boxing forward, (f) waving, (g) lying down, and (h) sitting up [31].

DCNN architectures based on the input features are listed below.

- *D*: This DCNN architecture involves feeding only the extracted Doppler spectrogram from a target into the 2D CNN. The distinctive micro-Doppler features in each activities are used by the DCNN to classify between different activities. In an example architecture, we have used 2D CNN with five convolutional layers followed by two fully connected layers.

- *RxD*: This DCNN architecture feeds in a video of RDI without further feature extraction into a 3D-CNN. As such, the video of RDI contains the most information, also capturing the spatio-temporal characteristics of the data, and thus to process such 3D-data require 3D CNN, which extends the input channels from 1 to the time length of the RDI sequence T. Correspondingly, the square filter kernels are transformed into cubic kernels applying separate weights across the spatio-temporal domain. Consequently, this architecture leads to a drastic increase in the number of parameters.

- *R+D late fusion*: In this architecture, both the input features, range spectrogram, and Doppler spectrogram are utilized by the DCNN architecture by means of late fusion. In the case of late fusion architecture, two independent branched convolutional layers are used to process features from both the range spectrogram and the Doppler spectrograms. The convolutional layers in both the branches have a similar structure to the D architecture. After the convolutional layers, the features from both streams are concatenated before feeding into the fully connected layers, whose dimensions are accordingly adjusted.

- *R+D Early Fusion*: In this architecture, both the input features (i.e., range spectrogram and Doppler spectrogram) are utilized by the DCNN architecture through feeding them as input channels and thus is referred as early fusion. In one of the implementations of early fusion, depth-separable convolutional layers (instead of standard convolution) can be used to process the distinct input feature images. In this case, the input features are fused much earlier in the DCNN and thus are referred to as early fusion. Depthwise separable convolution instead of normal convolution allows to achieve similar accuracy with a fewer number of parameters and FLOPs. It also prevents overfitting as the number of trainable parameters is reduced. The main idea behind depthwise separable convolution is that unlike normal convolution, first spatial convolution is applied to each channel separately and then a pointwise convolution (1×1 convolution) is applied that combines the intermediate channelwise output. Figure 4.6 illustrates how depthwise separable convolution works in comparison to normal convolution.

Figure 4.7 presents the high-level architecture of all the DCNN architectures for human activity classification. These DCNN architectures have issues due to time misalignment when they are deployed for drawing inference on a continuous stream of data. In a typical FMCW radar, a frame consists of a sequence of chirps followed by an off period wherein data is transferred to the host. Thus, if an activity starts within these off-periods would result in variations in input features due to shifted spectrogram start. Further, if there are missed detections in the first few frames or within an ongoing activity, recording can lead to variations. Such artifacts can lead to inaccuracies in online continuous activity classification. To avoid such issues, data augmentation is used by shifting training examples

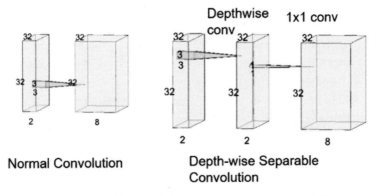

Figure 4.6   Depth-wise separable convolution compared to normal convolution.

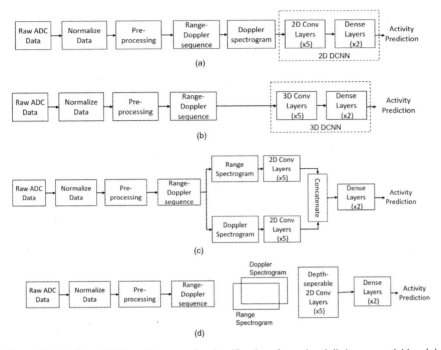

Figure 4.7 Various DCNN architectures for classification of everday daily human activities: (a) D, (b) RxD, (c) R+D late fusion, and (d) R+D early fusion.

toward the left and right to re-create the time misaligned input features typically experienced in real-life classification. Further, zeroing of inputs at arbitrary input locations to re-create the effect of a missed signal is applied to the training data. Such data augmentation during training helps in scaling the classifier for real-time artifacts and avoiding overfitting. Furthermore, for training, Adam optimization with a learning rate of 0.001 with cross-entropy loss function is an ideal optimization loss for this problem.

As well, for online classification, a sliding window can be deployed that extracts input features over consecutive time periods and is fed into the DCNN classifier. This is similar to cropping different image segments to the classifier, often with some transformations. This procedure is referred to as test-time augmentation and is an effective method in achieving scalability of the classifier in real-time inference. The final classification is generated through max-voting or median-voting of the classification result of these several input segments.

4.5.2 Results and Discussion

Table 4.1 outlines an example of the architecture of the four DCNN architectures involving about five convolutional layers followed by two dense layers. For the

Table 4.1

Architecture Details such as Output Size and Filter Kernel Size for all layers of: 1) D, CNN that takes Doppler spectrograms as input, 2) RxD, 3D-CNN accepting range-Doppler sequences as input, 3) R+D Late Fusion, multibranch-CNN for range and Doppler spectrograms, 4) R+D Early Fusion, depth separable CNN with range and Doppler spectrograms as input channels [10].

Layer	Output size				Filter Edge Size
	D	RxD	R+D Late Fusion	R+D Early Fusion	
input	64x64	64x64x64	64x64 (x2)	64x64x2	
conv1	64x64x8	64x64x64x8	64x64x8 (x2)	64x64x8 (separable)	5
maxpool1	32x32x8	32x32x32x8	32x32x8 (x2)	32x32x8	
relu1	32x32x8	32x32x32x8	32x32x8 (x2)	32x32x8	
conv2	32x32x16	32x32x32x16	32x32x16 (x2)	32x32x16 (separable)	3
maxpool2	16x16x16	16x16x16x16	16x16x16 (x2)	16x16x16	3
relu2	16x16x16	16x16x16x16	16x16x16 (x2)	16x16x16	3
conv3	16x16x16	16x16x16x16	16x16x16 (x2)	16x16x16 (separable)	3
maxpool3	8x8x16	8x8x8x16	8x8x16 (x2)	8x8x16	3
relu3	8x8x16	8x8x8x16	8x8x16 (x2)	8x8x16	3
conv4	8x8x32	8x8x8x32	8x8x32 (x2)	8x8x32 (separable)	5
maxpool4	4x4x32	4x4x4x32	4x4x32 (x2)	4x4x32	5
relu4	4x4x32	4x4x4x32	4x4x32 (x2)	4x4x32	5
conv5	4x4x64	4x4x4x64	4x4x64 (x2)	4x4x64 (separable)	1
relu5	4x4x64	4x4x4x64	4x4x64 (x2)	4x4x64	5
(flatten)	1024	4096	2048 (concat)	1024	
fc6	256	256	256	256	
relu6	256	256	256	256	
fc7	8	8	8	8	
softmax					

sake of comparison, the parameters of the four DCNN architectures have been chosen to be relatively similar to highlight the advantages and drawbacks of each architecture. The table outlines the input feature dimension at each layer and the edge size of filter kernels. ReLUs, the standard activation function, is applied after all layers except for the last layer where softmax is applied for generating the classification scores. 2×2 max-pooling is used to reduce height and width after the convolutional layers 1 to 4.

The data set has been split into a training set of 1,728 examples and a validation set of 384 examples. This data splitting has been performed randomly and homogeneously once for all architectures, so that every class is equally divided into training and test set. Figure 4.8 presents the confusion matrix of the different architectures on the test data set. The overall accuracy of D, RxD, R+D late Fusion, and R+D early fusion architectures are 91.87%, 93.25%, 92.75%, and 93%, respectively. As can be observed, the R+D fusion architectures result in accuracies

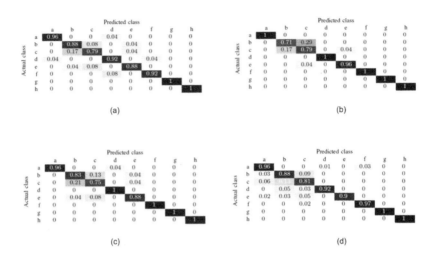

Figure 4.8 Confusion matrix of the test data set from different DCNN architectures for classification of everday daily human activities. Matrix keys: a = sitting, b = walking, c = running, d = boxing still, e = boxing forward, f = waving, g = lying down, and h = sitting up [10].

very close to that of the 3D DCNN with range-Doppler input data with very few parameters. The most common pairs of misclassification are between walking (b) and running (c), running (c) and boxing forward (e), and boxing still (d) and waving (f). Some of these misclassifications are due to minimal supervision provided to different participants executing these activities while recording for the training and test data set. From the accuracy results, it is evident that a range spectrogram brings in advantages in classification of some resembling activities. Also from the summary plot in Figure 4.9, it is evident that R+D early fusion and R+D late fusion are excellent candidates for accuracy vs. parameter compromise. In particular, R+D early fusion implemented using depth-separable convolutions can help reduce the parameters substantially while not compromising on the accuracy.

4.6 Bayesian Classification

The typical state-of-art radar processing pipeline, involves a detector followed by a tracker module, which includes the track-to-detect association, track filtering, and track management, which is further followed by extraction of features such as Doppler spectrogram for classification of target attributes. In the case of the constant velocity model, the tracker state parameters comprises the target position and velocity. However, continuous indoor human sensing and classification have challenges, such as

1. Multipath reflections from static targets such as walls, and furniture;

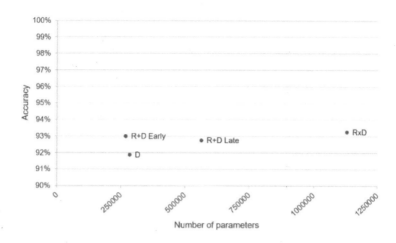

Figure 4.9   Accuracy vs. parameters for different DCNN architectures.

2. Ghost targets due to interfering nonhuman targets such as curtains, and ventilators;

3. Varying range-Doppler spread of human reflections;

4. Superposition of Doppler spectrogram due to closely spaced humans;

5. Unknown or transiting activity motions.

Some of these artifacts are depicted in Figure 4.10 and are the main source of misclassifications in practical deployed scenarios. To circumvent several of these challenges, Bayesian classification can be leveraged. One easy approach to achieve Bayesian classification is to integrate the classifier's output probabilities into the tracker state parameters. Thus, the classifier's output probabilities are then no longer a function of the classifier's output due to the current extracted Doppler spectrogram, which could be erroneous due to various artifacts as outlined earlier, but is also dependent on the state's output probabilities and their state transition matrix. The final classification is read out from the tracker's updated state parameter, which comprises the augmented classification probabilities. Likewise, the track-to-detect association, which is achieved through an ellipsoidal gating function, in the augmented state the track-to-classification association is also achieved through the ellipsoidal gating function. Thereby, the gating for classification would remove the outlier classification response from the classifier and state update of the activity probabilities happens only when the classification probabilities lie within the ellipsoidal gating. The unknown process model for the state transition function of the state activity probabilities can be calculated through supervised learning through a fully connected layer and is described in detail later. The Bayesian framework [30] compared to the state-of-the-art framework is presented in Figure 4.11.

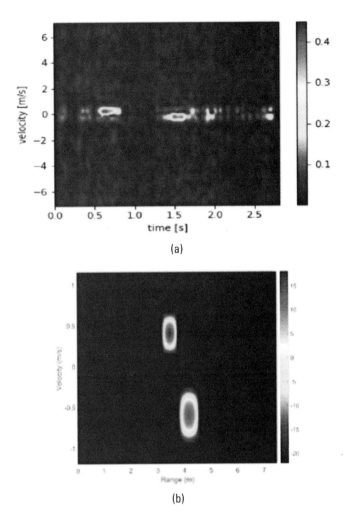

Figure 4.10 Artifacts due to the superposition of Doppler spectrogram in the same range bin, and activity-specific range-Doppler variation spread detections. (a) Two people in the same range bin, and (b) varying range-Doppler spread due to activities.

The Bayesian framework of integrating the classifier's activity probabilities into the tracker's state parameter improves the classification accuracy by smoothing several misclassifications arising due to the earlier mentioned artifacts. Further, since the Kalman filter based tracker not only provides the state estimation but also provides the associated uncertainty at each time step, it also outputs the uncertainty or confidence associated with an output activity class prediction, offering the system the flexibility to reject an activity prediction if the uncertainty associated with the predicted class is larger than a prespecified threshold.

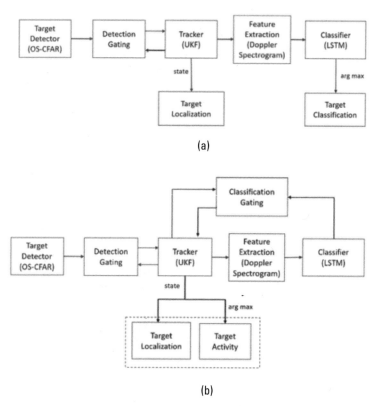

Figure 4.11   (a) Standard state-of-the-art algorithm pipeline of detector, followed by tracker and then classifier. (b) Proposed algorithm pipeline of detection followed by integrated tracking and classifier [30].

4.6.1   Integrated Classifier and Tracker

After the OS-CFAR and clustering through DBSCAN, the list of targets with their corresponding parameters, such as range, velocity, and angle, are fed into the tracker. These measurements are then fed into the tracker module, which updates its state parameter with the new measurement every frame provided it is within the ellipsoidal gating window and following the track-to-measurement association, such as probabilistic data association filter or suboptimal nearest neighbors. From the detected target's range, the Doppler spectrogram is extracted over 16 consecutive frames and fed into the classifier, whose classification probability is fed into the tracker module to update its augmented state parameters following likewise the ellipsoidal gating function. Alternately, the classification could be executed through a sliding windowed Doppler spectrogram over 16 consecutive frames. Due to the augmented state parameters, the overall process model and the measurement model are highly nonlinear and thus an unscented Kalman filter is required to approximate the nonlinear transformation from one state to another during the state prediction step and state domain to measurement domain during the state update step.

4.6.1.1 Augmented State Parameter

Based on the constant velocity model, the state vector is defined as

$$x = \begin{bmatrix} px & py & vx & vy \end{bmatrix}^T \tag{4.12}$$

where px, py, vx, py are the position coordinates and velocity values along the x- and y-axes, respectively. The classifier outputs the activity probabilities for a provided Doppler spectrogram, which is used to augment the state vector x^a as follows:

$$x^a = \begin{bmatrix} px & py & vx & vy & a^1 & a^2 & \cdots & a^M \end{bmatrix}^T \tag{4.13}$$

where a^i are the probability of class i and M is the total number of classes considered. The augmented covariance matrix P for the state vector x at any time step k thus not only takes account the uncertainty associated with the positional coordinates of the target but also that of the output probabilities of each class.

4.6.1.2 Augmented Process Model

The process model based on the constant velocity model is given as

$$x_{k+1|k} = \begin{bmatrix} 1 & 0 & \delta t & 0 \\ 0 & 1 & 0 & \delta t \\ 0 & 0 & 1 & 0 \\ 0 & 0 & 0 & 1 \end{bmatrix} x_{k|k} \tag{4.14}$$

where $k|k$ is added as a suffix to the state variable to denote the state update at kth time step using measurement at kth, whereas $k+1|k$ in the suffix indicates only the state prediction has been applied at time step $k + 1$. In the case of the augmented state parameters, the state transition/process function can be provided as

$$\hat{a}^i = \sum_{j=0}^{M} w_{ij} a^j$$

$$a^i{}_{k+1|k} = \frac{e^{\hat{a}^i}}{\sum_{i=0}^{M} e^{\hat{a}^i}} \tag{4.15}$$

where w_{ij} are the transition weights associated with transition of activity a^j to a^i, which is depicted in Figure 4.12. The second step of the process model is the softmax operation, which helps to generate the normalized state probability for activity class i, since first step in (4.15) does not guarantee that the activities class has a probability distribution after the state prediction step. This is a nonlinear

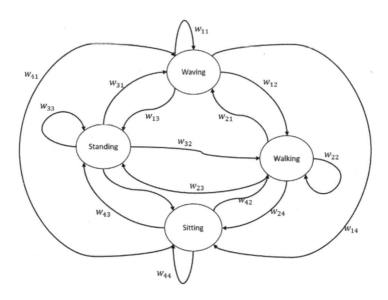

Figure 4.12 Representation of the state transition of augmented state parameters.

operation, thus the process model can be expressed as

$$x^a_{k+1|k} = g(x^a_{k|k}) \tag{4.16}$$

where $g(.)$ contains both the linear and nonlinear part given by (4.14) and (4.15) and thus is overall nonlinear.

4.6.1.3 Augmented Measurement Model

The measurement model accounts for the transformation of the state vector into the measurement domain, where the innovation is applied with the Kalman gain. The measurement model for the unaugmented state parameters are given as

$$\rho_{k+1|k} = \sqrt{(px_{k+1|k})^2 + (py_{k+1|k})^2}$$

$$v_{k+1|k} = \frac{px_{k+1|k} \cdot vx_{k+1|k} + py_{k+1|k} \cdot vy_{k+1|k}}{\sqrt{(px_{k+1|k})^2 + (py_{k+1|k})^2}}$$

$$\theta_{k+1|k} = \tan^{-1}\left(\frac{py_{k+1|k}}{px_{k+1|k}}\right) \tag{4.17}$$

where $\rho_{k+1|k}$, $v_{k+1|k}$, $\theta_{k+1|k}$ are the predicted radial range, radial velocity or range-rate, and azimuth angle, respectively. For the augmented parameters (i.e., the classification probabilities), this is an identity mapping from the state domain to the measurement domain. However, the overall measurement model is a nonlinear transformation $\hat{z}_{k+1|k} = h(x_{k+1|k})$.

4.6.1.4 Unscented Transformation

The nonlinear process model and the measurement model can be achieved through unscented transformation using sigma points generated to approximate the statistical distribution of the state parameter from one domain to another. Considering a random vector η, which is Gaussian distributed with mean μ and covariance Ω, on performing nonlinear transformation $\psi = \phi(\eta)$ would lead to another Gaussian distribution. The unscented transform to approximate this nonlinear function is done through generating a set of fixed sigma points χ drawn from the initial distribution as

$$\chi^{(0)} = \mu$$

$$\chi^{(i)} = \mu + \sqrt{\left(\frac{n}{1-W_0}\right)}\Omega_i^{\frac{1}{2}}, i = 1, 2, ..., n_\eta$$

$$\chi^{(i+n)} = \mu - \sqrt{\left(\frac{n}{1-W_0}\right)}\Omega_i^{\frac{1}{2}}, i = 1, 2, ..., n_\eta \qquad (4.18)$$

where $W_i = \frac{1-W_0}{2n}$ and $\Omega_i^{\frac{1}{2}}$ is the ith column of $\Omega^{\frac{1}{2}}$, which is the Cholesky decomposition of matrix Ω. In total, $2n_\eta + 1$ sigma points are generated with n_η being the dimension of the state η and in our case is $4 + M$.

Since the nonlinear transformation of a Gaussian distribution is also a Gaussian distribution, they are fully characterized by their mean and covariance. The nonlinear transformation can be effected by calculating the mean and covariance of the nonlinear transformed Gaussian distribution as

$$E[\psi] \approx \sum_{i=0}^{N} W_i \phi(\chi^{(i)})$$

$$Cov\{\psi\} \approx \sum_{i=0}^{N} W_i (\phi(\chi^{(i)}) - E[\psi])(\phi(\chi^{(i)}) - E[\psi])^T \qquad (4.19)$$

4.6.1.5 Process Transition Function

As an example, the classifier used to predict the activities based on the extracted Doppler spectrogram can be realized either through DCNN or LSTM. The DCNN or LSTM is first trained to classify the correct activities through cross-entropy loss function and supervised training of sequence of the Doppler spectrum of the activity. The backpropagation-through-time model of the LSTM during training and also during inference is presented in Figure 4.13.

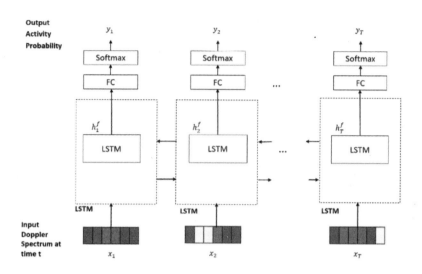

Figure 4.13 LSTM classification and backpropagation-through-time learning through sequence of Doppler spectrum.

Then, secondly, for learning the weights of the transition matrix of the activity probabilities, a fully connected layer followed by a softmax layer is attached at the output of the trained LSTM classifier. The weights of the fully connected layer w_{ij} are learned during this second training step by feeding a sequence of Doppler spectrograms that contains the transitions between activities, while keeping the weights of the classifier fixed. This means unlike in the first step where the labeled data contained only a single activity, the second training step includes labeled data containing pair of activities, such as walking to sitting, sitting to waving, and so on. The training procedure of the second step is described in Figure 4.14. Once the weights in the fully connected layer are learned, they are utilized in the UKF state transition function for transformation of the augmented parameters as mentioned in (4.15).

4.6.2 Results and Discussion

For the Bayesian framework-based solution for continuous human activity classification, we considered four different activities; walking, standing, sitting, and walking. Figure 4.15 presents an example of continuous Doppler spectrogram wherein a human executes the four different activities and is sensed by *Infineon's BGT60TR13C* 60-GHz FMCW radar sensor. The chipset is configured for a bandwidth of 1 GHz resulting in a range resolution of 15 cm, pulse repetition time (i.e., time between start of two chirps), as 250 μs with a chirp time of 37 μs. The number of ADC samples are 128 with 64 chirps in a frame, resulting in a

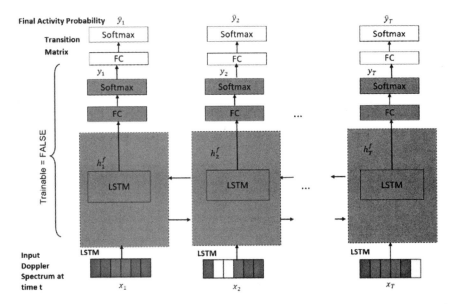

Figure 4.14 LSTM parameters are frozen and output is connected to a fully connected layer, and the additional fully connected layer is trained to learn the transition probabilities.

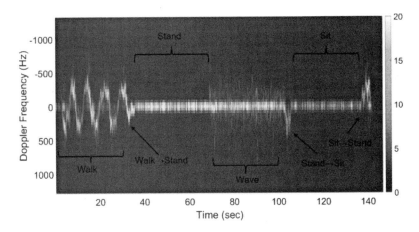

Figure 4.15 Example Doppler spectrogram of continuous human activities indicating different activities and also the transitions among activities [30].

Doppler resolution of $0.3125\ m/s$. The frame time is set to 75 ms, resulting in a frame rate of 13.33 Hz.

To demonstrate the framework, we trained a LSTM model with a 64-dimension Doppler spectra vector as input and having 128 hidden states and was trained to predict the activity label after 16 consecutive Doppler spectrogram of the target are fed in for a particular tracked target. The training data set was collected with the help of five individuals from different rooms and has 800 Doppler spectrograms of the four activities in total. The trained LSTM model achieved a classification accuracy of 87.24% on the test data set, which comprised 200 Doppler spectrograms of the different activities. As explained in the process transition function, for learning the transition matrix of the activity probabilities, we add a fully connected layer along with a softmax layer to the trained LSTM model. To learn the weights of the fully connected layer, which will be eventually used in the tracker during inference, supervised training examples of Doppler spectrograms are fed as input capturing the transitions between activity pairs. A set of 1,000 training examples were used to learn the transition function for the activity probabilities, which are generated by the trained LSTM model. The detected target's parameters are fed into the UKF tracker every frame, whereas the classification probabilities are updated every sixteenth frame. However, an update of the classification probabilities every frame is also possible through sliding window classification using the same classifier. The final classification of the activity at any sixteenth frame is done through *argmax*() (i.e., picking the class with maximum probability from the tracker's state). In the non-Bayesian setting, the *argmax*() is applied directly on the LSTM classifiers output.

A comparison of ground truth on an example continuous activity, along with a corresponding prediction from the non-Bayesian approach and the Bayesian approach, is provided in Figure 4.16. It can be observed that the classifier predicts several incorrect classes especially between the transition of activities such as standing and sitting. However, the Bayesian approach is able to remove such

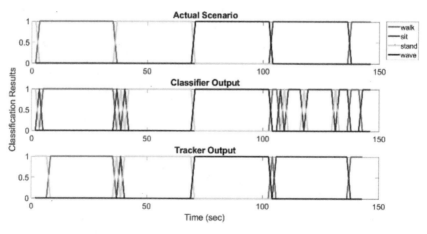

Figure 4.16 Top shows the actual scenario, the middle shows the classification output from the classifier, and the bottom shows the output of the activity states after the combining processing with the the tracker and classifier [30].

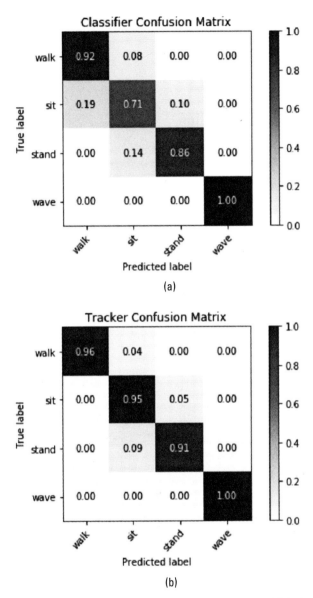

Figure 4.17 (a) Confusion matrix for the output of the conventional LSTM classifier, and (b) confusion matrix for the proposed solution over the same test data set [30].

incorrect classifications and is able to considerably stabilize the output results close to the ground truth. Furthermore, the overall confusion matrix on the test data set using the non-Bayesian and Bayesian approach is presented in Figure 4.17. The integration of the tracker is able to improve the classification accuracy of the LSTM classifier from 87.24% to 95.5%.

Further, the tracker solution also generates the uncertainty associated with the activity probabilities at each iterative update step, which can be further utilized for reducing false alarms during the transitions between activities and Doppler spectrogram from unknown activities.

4.7 Fall-Motion Recognition

In contrast to human activity classification, identification and classification of the fall motion is a different and rather challenging problem since fall motion can be equated to an anomaly class, which differs drastically from the usual activity class and is sufficiently rare with a unexpected data distribution. Broadly, anomaly detection can be achieved either through unsupervised techniques, where in the anomaly class is identified through the assumption that it does not fit into the majority of the data, or the supervised mechanism, which requires normal and abnormal data labels for classification. For the supervised setting, the problem of class imbalance needs to be dealt with since the data set most likely would comprise a large set of normal human motions with much less data for fall motions. If not accounted for in the design of the Deep Net architecture, such class imbalances can lead to degenerate models, wherein training loss is quite low but the model is biased toward the normal class and does not generalize. A common solution is to perform hard negative mining that samples hard examples during training, such as synthetic minority oversampling technique (SMOTE); however, such techniques do not usually scale very well for deep neural networks. Thus due to the problem construct, fall-motion detection is an independent problem outside of the activity classification problem. Furthermore, fall-motion recognition using radars has several other challenges that need to be addressed for practical deployment of such systems [31]:

- First, the inconspicuous interclass difference between true fall-motion Doppler spectrogram and other closely related motions, such as bending down to pick up an object from the ground, can be visibly close to that of a fall motion. Generating a classifier considering all the variations is practically not possible solely due to the wide possibility of closely resembling a Doppler spectrogram with that of an actual fall motion.

- Second, there is a large intraclass variation of different types of fall motion, namely loose balance fall, loss of consciousness fall, trip fall, slip fall, reach, and fall [21], which have quite distinctive Doppler spectrograms.

- Third, a radar sensor can be placed anywhere in the room (that is, the sensor could sense the falling motion when the person is directly facing the sensor, or when the falling subject is falling behind or at the side or at oblique angles to the sensor placement. Figure 4.18 presents the Doppler

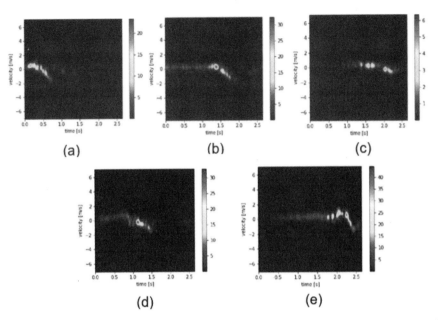

Figure 4.18   2D Doppler spectrograms of various fall detection sensed by the radar placed behind the falling subject near the ceiling with a shearing angle toward the falling subject: (a) loose balance fall, (b) loss of consciousness fall, (c) reach fall, (d) slip fall, and (e) trip fall [31].

spectrogram of various fall types when the subject fell behind and the radar was mounted on the ceiling with a shearing angle. Figure 4.19 presents the same Doppler spectrogram of the fall-motion sensing the falling subject in front at the same height whereas Figure 4.20 presents the Doppler spectrogram, where the radar sensed the falling subject from the side at the same height. Since the radar senses the radial Doppler velocities it is clear that the Doppler prominence in Figure 4.20 are less comparable to that of Figures 4.18 and 4.19.

Such practical challenges can be addressed by proper consideration of these factors in designing the neural network. The inconspicuous-looking Doppler spectrogram from interclasses can be addressed through the one-class contrastive loss function, which is inspired by the general contrastive loss [25] similar to one proposed in [26]. Further, the problem of the same Doppler spectrogram being highly dependent on the aspect angle of the sensed human target can be addressed through the deformable convolutional neural network [27], where these nonlinear transforms are learned through supervised training via local, nonparameteric convolution/RoI pooling modules. The deformable convolution layers and deformable pooling layers along with the one-class contrastive loss can generate embedding vectors closely spaced in the latent space of fall motion from

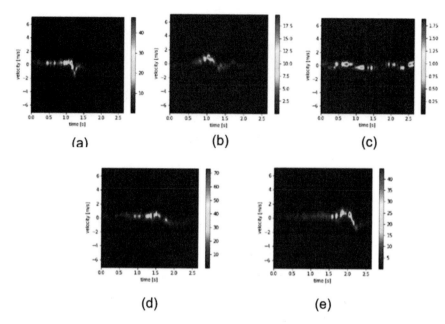

Figure 4.19   2D Doppler spectrogram of various fall detection sensed by the radar placed in front of the falling subject: (a) loose balance fall, (b) loss of consciousness fall, (c) reach fall, (d) slip fall, and (e) trip fall [31].

the above sources of variation (i.e., aspect angle, interclass, and intraclass). Further, the issue of training a neural network with class imbalance can be effectively dealt with the focus loss function [28], which can adaptively prioritize hard examples over soft examples during training through a simple modification of the cross-entropy loss function. Hard examples are the ones that belong to the class and in the learned embedding space during ongoing training phase are close to the incorrect class's cluster.

4.7.1   Architecture and Learning

As mentioned earlier, since the Doppler spectrograms for different fall motion exhibits variations due to the aspect angle of the falling subject and also due to intraclass variation. Such Doppler spectrogram variation sensed by the radar is a nonlinear transformation that is not based on fixed affine geometric transformation and hence is required to be learned. To address such challenges, deformable (adaptive) convolution and deformable (adaptive) RoI pooling can be used instead of conventional (nonadaptive) convolution and (nonadaptive) RoI pooling. In deformable convolution and RoI pooling, each sampling pixel is biased by a learnable offset resulting in a deformed grid over which convolution and RoI pooling are performed. This, effectively allows the network to learn

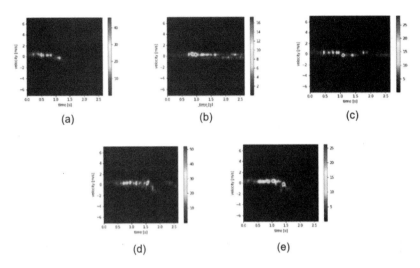

Figure 4.20 2D Doppler spectrogram of various fall detections sensed by the radar placed at the side of the subject: (a) loose balance fall, (b) loss of consciousness fall, (c) reach fall, (d) slip fall, and (e) trip fall [31].

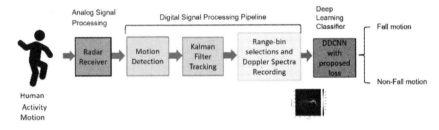

Figure 4.21 End-to-end processing pipeline of proposed nonintrusive elderly fall-motion detection system using 60-GHz BGT60TR24B radar chipset [31].

nonlinear transformations and adaptive global receptive fields. Similar nonlinear transformations could be learned through spatial transformer networks [29] or a T-network. The DDCNN architecture was first proposed in [27]. The end-to-end proposed processing pipeline is depicted in Figure 4.21.

4.7.2 Deformable CNN

An example architecture used for the fall-motion recognition problem using Doppler spectrogram is presented in Figure 4.22.

4.7.2.1 Deformable Convolution

Deformable convolution involves two parts: regular convolution that operates on a rectangular grid $\mathcal{R}$ and another convolution layer that operates on the deformed

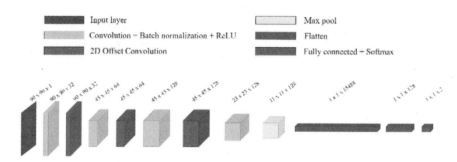

Figure 4.22   Example deformable DCNN with four deformable convolutional layers, a max-pooling layer followed by a fully connected layer. The data dimensions of the data flow over the network are also indicated [31].

grid formed by augmenting points of $\mathcal{R}$ with learnable 2D offset δp_n. The 2D offsets (x-direction and y-direction for each offset) are encoded in the channel dimension and convolutions generates 2N feature maps for N offsets.

- Regular convolution

$$y(p_0) = \sum_{p_n \in \mathcal{R}} w(p_n) x(p_0 + p_n) \tag{4.20}$$

- Deformable convolution

$$y(p_0) = \sum_{p_n \in \mathcal{R}} w(p_n) x(p_0 + p_n + \delta p_n) \tag{4.21}$$

where $w(p_n)$ are the convolution weights, $\mathcal{R}$ is the learned receptive field and δp_n are the offsets generated by a sibling branch of regular convolution.

4.7.2.2 Deformable RoI Pooling

Deformable RoI pooling also involves two parts: first, a regular RoI pooling to generate a pooled feature map, and a fully connected layer that learns the offset. Instead of predicting the raw offset, the offsets are normalized to make it invariant to RoI size by dividing it with the width and height of the RoI region. Thus, the fully connected layer yields normalized offsets δp_{ij} which are transformed as given in (4.23). Finally, an output feature map is generated by pooling on augmented regions.

- Regular RoI pooling

$$y(i,j) = \sum_{p \in bin(i,j)} x(p_0 + p)/n_{ij} \tag{4.22}$$

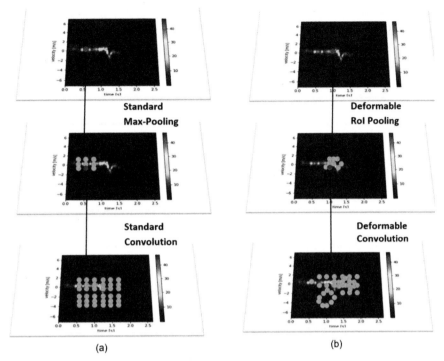

(a)   (b)

Figure 4.23 Illustration of the difference between sampling locations for (a) standard convolution, where locations are fixed, and (b) deformable convolution, where locations are learned during training [31].

- Deformable RoI pooling

$$y(i,j) = \sum_{p \in bin(i,j)} x(p_0 + p + \delta p_{ij})/n_{ij}, \qquad (4.23)$$

where δp_{ij} is generated by a sibling fully connected branch.

The differences between standard convolution with max-pooling and deformable convolution with associated max-pooling are depicted in Figure 4.23.

4.7.3 Loss Function

The loss function for this problem has two major objectives; one is to utilize the descriptive features required for classification (i.e., descriptive loss), and the other is to project the examples in the latent feature space, so that the fall-motion classes are close together (i.e., compactness loss). The effective loss combines the descriptive loss and the compactness loss. This helps not only in classifying the binary classes but also by projecting the data into an embedding space wherein all fall-motion data appear clustered together closely. This helps in the model to generalize to fall motions in alien conditions since they would be projected close to the fall cluster compared to nonfall cluster in the embedding space. Likewise, the same applies for unseen nonfall activities not part of the initial training data set. A weighted combination of the following losses are used

4.7.3.1 Descriptive Loss

The descriptive loss can be achieved through binary cross entropy while applying a focal loss in order to address the class imbalance issue. The binary cross-entropy (CE) loss for binary classification is given as

$$L_{CE(p,y)} = -y \log(p) - (1 - y) \log(1 - p) \tag{4.24}$$

However, the problem with binary cross entropy in its original form is that the fall class or anomaly class can be completely neglected over the easy examples of the nonfall class, thus leading to a degenerate model. Such issues can be addressed through focal loss, where the original binary cross-entropy loss can be modified through γ parameter as defined as

$$L_{FL}(p, y) = -y(1 - p)^\gamma \log(p) - (1 - y)p^\gamma \log(1 - p) \tag{4.25}$$

When an example is misclassified and p is small, the modulating factor is near 1 and the loss is unaffected. As $p \rightarrow 1$, the factor goes to 0 and the loss for well-classified examples is down-weighted [28]. The focusing parameter γ smoothly adjusts the rate at which easy examples are down-weighted, thus preserving the cross-entropy loss alongside addressing the class imbalance issues.

4.7.3.2 Compactness Loss

The compactness loss in the latent embedding space can be addressed through the one-class contrastive loss, which attempts to project the fall-motion Doppler spectrogram data closer in the feature embedding space. One-class contrastive loss is quantified as the mean-squared fall-class distance between an example and its mean cluster in the embedding space. Mathematically, let $\{x_1, x_2, \cdots, x_n\} \in \mathcal{R}^{1 \times k}$ be the example data with a batch size of n during training, and then the one-class contrastive loss can be defined as

$$L_{CL} = \frac{1}{nk} \sum_{i=1}^{n} (x_i - m)^T (x_i - m) \tag{4.26}$$

where m is the mean of the fall-motion class. Therefore, the effective loss function is defined as

$$L(p, y) = -(1 - y)p^\gamma \log(1 - p) + \left(-y\alpha(1 - p)^\gamma \log(p) + y(1 - \alpha)L_{CL} \right) \tag{4.27}$$

where α is the hyperparameter to control the proportion of contrastive loss and the contribution of focal loss for fall class toward the overall effective loss.

4.7.4 Results and Discussion

A training data set with 1,244 records with 302 staged fall motions (general falling, loss of consciousness fall, balance loss fall, etc.) and 942 nonfall motions (sitting, walking, walking with a stick, waving, running performed in forward, backward, and sideways, etc.) were collected. The fall motion was staged and performed by healthy adults for data collection. The test data contains 600 nonfall motions spectrograms and 200 fall motion spectrograms. To evaluate the generalization capability of the model, the test data had examples, such as fall motion sensed by radar placed obliquely to the falling subject and nonfall motions such as exercises (bend and rise) and roll sideways, etc. which were not available during training. An example of the spectrograms from unseen fall and nonfall motion included in the test data are depicted in Figure 4.24. Once detected, the Doppler spectrogram was captured for 90 consecutive frames, leading to a dimension of 128×90, which was cropped to 90×90 since larger Doppler events rarely were seen in the data.

The DDCNN was trained wherein the convolutional layers were initialized with values drawn randomly from normal distribution with zero mean and a standard deviation of 0.01 whereas the bias was drawn from normal distribution with 0.5 mean instead of 0. The dense layer weights are initialized by sampling from a Xavier uniform distribution. For the effective loss function, the focusing parameter γ was set to 2 while α was set to 0.5. An Adam optimizer was used to update the weights during backpropagation. Further, to avoid overfitting of the

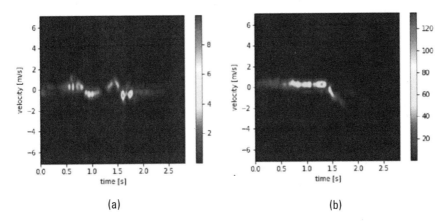

(a)   (b)

Figure 4.24   New activity from an exercising human and fall motion sensed by the radar at a oblique angle to the falling subject are used in the test data set not present in the training data set.

Table 4.2
Comparison of Different Convolutional Neural Network, Along with
Different Loss Functions [31]

Approach	Description	Accuracy (%)
DCNN with CE	Standard convolution layers with cross-entropy loss	96.33
DCNN with EL	Standard convolution layers with proposed effective loss	98
DDCNN with CE	Deformable convolution layers with cross entropy	98.67
DDCNN with EL	Deformable convolution layers with proposed effective loss	99.5

network, we applied dropout to the first and third convolutional layers. Dropout of 0.25 and 0.5 was applied to the first and third convolutional layers.

Table 4.2 compares the different architectures of the same dimensions along with different loss functions. Compared to normal DCNN, DDCNN with effective loss function was able to generalize to unseen fall motion and nonfall motion. The t-SNE visualization of two DDCNN models with and without effective loss function; that is, focus loss in conjunction to one-class contrastive loss, is presented in Figure 4.25 and demonstrate that the effective loss function was able to better learn clustered representation in the embedding space, thus generalizing better compared to its counterpart. Overall, the DDCNN architecture with effective loss attained an accuracy of 99.5%.

4.8 Future Work and Directions

There are still several open challenges of activity and fall-motion classification. A unified system combining both the activity and fall motion as an anomaly class designed through meta-learning is an open problem. This solution requires quadruplate distance-metric loss, where the anomaly fall class needs to have a higher margin compared to normal negative classes from normal activities. As well, for everyday activities, the Doppler spectrogram from a human activity is perturbed by another human targets activity in multihuman environment and also other moving objects, such as a fan, or ventilator in the room, causing incorrect predictions. Occurrence and denoising of such super-imposed Doppler spectrograms isn't possible through conventional signal processing. In such cases, the Doppler spectrogram has a distorted signature compared to what the model was trained on, and thus deep domain adaptation techniques need to deployed to address such variations during deployment. Furthermore, some activities such as sitting or standing still have inconspicuously similar Doppler spectrogram

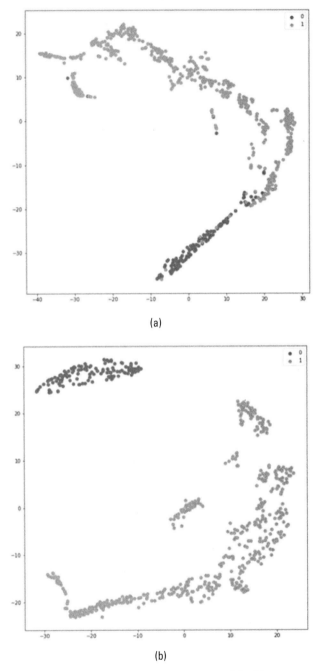

(a)

(b)

Figure 4.25 t-SNE representation of fall-motion embedding: (a) without custom loss, and (b) with custom loss, demonstrating that the fall-motion data generates a close-knit cluster in the embedding space before classification hyperplane [31].

signature, thus to distinguish them requires additional data or sensor sources, which could be a future research direction.

4.9 Problems

1. What is the need for tracking the target range spread length l_k and its rate of change $\dot{l}_k$ for activity classification using a Doppler spectrogram?
2. What are the limitations of range a spectogram alone for activity classification?
3. Define depthwise-seperable convolution.
4. Compare early-fusion and late-fusion architectures for utilizing range spectrograms and Doppler spectrograms for classification.
5. How does the classification gating help the continuous activity classification solution?
6. Why is unscented transformation needed for tracking the activity class probabilities?
7. What is deformable convolution? How does deformable convolution (compared to normal convolution) help in the context of radar-based activity or fall classification?
8. What is 1-class contrastive loss and why does it help in anomaly classification in comparison to binary cross entropy?
9. How many additional parameters does deformable convolution introduce compared to normal convolution?
10. How can deformable convolution be implemented on a HW accelerator with the provision of performing only convolution operation?

References

[1] Chen, V. C., *The Micro-Doppler Effect in Radar*, Norwood, MA: Artech House, 2011.

[2] World Population Prospects: The 2010 Revision, United Nations, Department of Economic and Social Affairs, Population Division, 2011, http://www.esa.un.org/wpp/.

[3] Molchanov, P., A. Totsky, J. Astola, et al., "Aerial Target Classification by Micro-Doppler Signatures and Bicoherence-Based Features," *European Radar Conf.*, 2012, pp. 214–217.

[4] Li, Y., L. Du, and H. Liu, "Hierarchical Classification of Moving Vehicles Based on Empirical Mode Decomposition of Micro-Doppler Signatures," *IEEE Trans. Geosci. Remote Sens.*, Vol. 51, No. 5, 2013, pp. 3001–3013.

[5] Anderson, M., and R. Rogers, "Micro-Doppler Analysis Of Multiple Frequency Continuous Wave Radar Signatures," *SPIE Proc. Radar Sensor Technology*, May 2007, Vol. 6547.

[6] Geisheimer, J. L., E. F. Greneker, and W. S. Marshall, "High-Resolution Doppler Model of the Human Gait," *SPIE Proc. Radar Sensor Technology and Data Visualization*, Vol. 4744, July 2002, pp. 8–18.

[7]   Otero, M., "Application of Continuous Wave Radar for Human Gait Recognition," *SPIE Proc., Signal Processing, Sensor Fusion, and Target Recognition XIV*, Vol. 5809, May 2005, pp. 538–548.

[8]   Ram, S., C. Christianson, Y. Kim, and H. Ling, "Simulation and Analysis of Human Micro-Dopplers in Through-Wall Environments," *IEEE Trans. Geosci. Remote Sens.*, Vol. 48, 2010, pp. 2015–2023

[9]   Yang, Y., and C. Lu, "Human Identification Using Micro-Doppler Signatures," *Proc. ARP*, 2008, pp.69–73.

[10]   Hernangómez, R., A. Santra, S. Stańczak, "Human activity classification with frequency modulated continuous wave radar using deep convolutional neural networks." In 2019 International Radar Conference (RADAR), pp. 1-6. IEEE, 2019.

[11]   Fairchild, P. D., and M. R. Narayanan, "Classification and Modeling of Human Activities Using Empirical Mode Decomposition with S-Band and Millimeter-Wave Micro-Doppler Radars," *SPIE Radar Sens. Technol. XVI*, Vol. 8361, Article No. 83610X, 2012, pp. 115.

[12]   Kim, Y., and T. Moon, "Human Detection and Activity Classification Based on Micro-Doppler Signatures Using Deep Convolutional Neural Networks," *IEEE Geoscience and Remote Sensing Letters*, Vol. 13, No.1, Jan 2016.

[13]   Zwart, J., R. Heiden, S. Gelsema, and F. Groen, "Fast Translation Invariant Classification of HRR Range Profiles in a Zero Phase Representation," *IEE Proc Radar Sonar Navigat*, Vol. 150, No. 6, 2003, pp. 411–418.

[14]   Xing, M.-D., Z. Bao, and B. Pei, "The Properties of High-Resolution Range Profiles," *OptEng*, Vol. 41, No. 2, 2002, pp. 493–504.

[15]   Du, L., H.-W. Liu, and Z. Bao, "Radar HRRP Target Recognition Based on Higher-Order Spectra," *IEEE Trans Signal Process*, Vol. 53, No. 7, 2005, pp. 2359–2368.

[16]   Liu, L., M. Popescu, M. Skubic, M. Rantz, T. Yardibi, and P. Cuddihy, "Automatic Fall Detection Based On Doppler Radar Motion," in *Proc. 5th Int. Conf. Pervasive Computing Technologies for Healthcare*, Dublin, Ireland, May 2011, pp. 222–225.

[17]   Kim, Y., and H. Ling, "Human Activity Classification Based on Micro-Doppler Signatures Using a Support Vector Machine," *IEEE Transactions on Geoscience and Remote Sensing*, Vol. 5, No. 47, May 2009, pp. 1328–1337.

[18]   Tomii, S., and T. Ohtsuki, "Falling Detection Using Multiple Doppler Sensors," *Proc. IEEE Int. Conf. e-Health Networking, Applications and Services*, Beijing, China, October 2012, pp. 196–201.

[19]   Su, B. Y., K. C. Ho, M. J. Rantz, and M. Skubic, "Doppler Radar Fall Activity Detection Using the Wavelet Transform," *IEEE Transactions on Biomedical Engineering*, Vol. 62, No. 3, Mar. 2015, pp. 865–875.

[20]   Wu, Q. Y., D. Zhang, W. Tao, and M. G. Amin, "Radar-Based Fall Detection Based on Doppler Time-Frequency Signatures for Assisted Living," *IET Radar, Sonar & Navigation*, Vol. 9, No. 2, 2015, pp. 164–172.

[21]   Amin, M. G., Y. D. Zhang, F. Ahmad, and K. C. Ho, "Radar Signal Processing for Elderly Fall Detection," *IEEE Signal Processing Magazine*, Vol. 33, No. 2, 2016, pp. 71–80.

[22] Chen, Z., G. Li, F. Fioranelli, and H. Griffiths, "Personnel Recognition and Gait Classification Basedon Multistatic Micro-Doppler Signatures Using Deep Convolutional Neural Networks," *IEEE Geoscience Remote Sensing Letters*, Vol. 15, 2018, pp. 669–673.

[23] Klarenbeek, G., R. I. A. Harmanny, and L. Cifola, "Multi-Target Human Gait Classification Using LSTM Recurrent Neural Networks Applied to Micro-Doppler," in *Proceedings of the European Radar Conference*, Nuremberg, Germany, October 11–13,2017, pp. 167–170.

[24] Jokanovic, B., M. Amin, and F. Ahmad, "Radar Fall Motion Detection Using Deep Learning," *IEEE Radar Conference (RadarConf) 2016*, pp. 1–6.

[25] Koch, G., R. Zemel, and R. Salakhutdinov, "Siamese Neural Networks for One-Shot Image Recognition," in *ICML Deep Learning Workshop*, Vol. 2, July 2015.

[26] Perera, P., and V. M. Patel, "Learning Deep Features for One-Class Classification," *IEEE Transactions on Image Processing*, Vol. 28, No. 11, 2019, pp. 5450–5463.

[27] Dai, J., H. Qi, Y. Xiong, Y. Li, G. Zhang, H. Hu, and Y. Wei, "Deformable Convolutional Networks,"in *Proceedings of the IEEE International Conference on Computer Vision*, 2017, pp. 764–773.

[28] Lin, T. Y., P. Goyal, R. Girshick, K. He., and P. Dollar, "Focal Loss for Dense Object Detection," in *Proceedings of the IEEE International Conference on Computer Vision*, 2017, pp. 2980–2988.

[29] Jaderberg, M., K. Simonyan, A. Zisserman, and K. Kavukcuoglu, "Spatial Transformer Networks, in *Advances in Neural Information Processing Systems*, 2015, pp. 2017–2025.

[30] Vaishnav, P., and A. Santra, "Continuous Human Activity Classification with Unscented Kalman Filter Tracking Using FMCW Radar," *Sensors Letters*, 2020.

[31] Shankar, Y., S. Hazra, and A. Santra, "Radar-Based Non-intrusive Fall Motion Recognition Using Deformable Convolutional Neural Network," in *18th IEEE International Conference on Machine Learning and Applications (ICMLA)*, 2019, pp. 1717–1724.

5

Air-Writing

5.1 Introduction

Air-writing is an advanced user interface that enables writing linguistic characters virtually in three-dimensional open space through hand-motion gestures. Users can write text as if on an imaginary board. Such interfaces are convenient alternatives to the traditional mechanism of typing on a keyboard or writing on a trackpad/touchscreen. However, air-writing using radars presents a different problem to gesture recognition since character recognition in an air-writing system must utilize the temporal trajectory of the hand marker in 3D space. Air-writing systems, unlike conventional writing, have several challenges since users have to perform character gestures relying only on visual cues of imaginary axes in three-dimensional space. Further, such systems lack a reference position on the writing plane and thus the notion of imaginary start and end coordinates; additionally, such systems need to automatically detect the start and end of a handwritten character in air. These considerations pose several challenges, increasing the intraclass variability of writing patterns of a letter.

Most of the proposed character recognition systems in the literature use hand-wearable devices. In [4], an approach to mobile text entry for wrist-worn wearable systems is proposed. The authors used motion sensors to extrapolate air-written letters and then used dynamic time warping for letter classification. A similar type of work is proposed in [4], using inertial sensors attached to the hand. The method is capable of spotting and continuously recognizing text written in the air. The main disadvantages of the wearable-based system is that sometimes the wearable devices are quite expensive and are also cumbersome to the users. In [6], handwritten text in an air-recognition system is proposed that first segments the words written by users in the air with the help of a Leap motion controller and then recognizes them using a hidden Markov model classifier. Zhang et al. [7] proposed a system for air-writing by combining depth and hand skin color information using a kinetic sensor. They used a depth-skin-background mixture model and artificial neural network for segmentation of the hand. After hand segmentation, a dual-mode (side and frontal) switching algorithm for fingertip detection and tracking is proposed. Finally, a compact modified quadratic discriminant function character classifier is applied for finger-writing trajectory recognition. In [8], a multilanguage unistroke air-written numeral recognition system has been proposed using DCNN.

This chapter proposes an air-writing system on a virtual board using a network of millimeter wave FMCW radars. A stroke is an isolated writing trajectory written in the air involving a sequence of hand motions. To estimate the correct trajectory of the gesture, the sensing system requires accurate hand marker localization. Trilateration techniques are employed to accurately estimate the hand marker in three-dimensional coordinates and smoothing filter to track the trajectory of the hand motion. For the subsequent recognition stage to generate the text representation from the hand motion sensor data, we explore LSTM and its variants with CTC loss function and DCNN for character recognition. For LSTM-CTC, the smoothed target trajectory is fed as input over time, whereas in the case of DCNN a 2D image is first reconstructed from the trajectory before being fed into the DCNN. LSTM with CTC loss function has the advantage that it facilitates continuous writing of characters to form a word, whereas in the case of DCNN an explicit indication of the end of a character has to be signaled to the system by removing the hand marker from the virtual board.

As another approach, we feed the range trajectory image from each radar into a 2D CNN to act as feature extractor and is used for modeling the change of ranges across time from the three radar sensors. It can be shown that whereas in the approach with trilateration a mimumum of three radars are required, but in the case of 2D CNN with range trajectory images as input, the character recognition can be achieved through a single radar range trajectory image too. As yet another approach, a 1D CNN-LSTM solution is proposed where range data from the three radars are passed as channel and the trilateration algorithm is removed and functionally executed by the 1D CNN. The advantage of the approach is that the solution is then able to operate in the case of sparse radar configuration (i.e., one or two radars), thus is able to deal with shadowing or occlusion, which would cause the trilateration-based recognition solution to fail since it requires data from atleast three radars in the network.

This chapter is laid out as follows: Section 5.1 presents the topic of radar network placement to maximize the intersection of a radar's field of

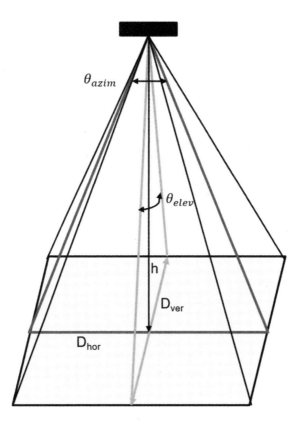

Figure 5.1   Radar FOV when radar is placed without any sheering angle. The gray patch indicates the FOV at height *h*.

view. In Section 5.2, we present radar preprocessing to extract a smooth target trajectory for each radar in the network, followed by trilateration to either generate a 2D trajectory image or time-series data. Section 5.3 presents the setup of radar networks and characters that were collected using three *Infineon's BGT60TR24B* radar sensor in a network. Section 5.4 presents a LSTM-based solution for character recognition comparing the classification accuracy from different variants, Section 5.5 presents DCNN-based solution for character identification, and Section 5.6 presents another solution using 1D CNN-LSTM, where the triangulation algorithm has been removed and is approximated through the 1D CNN.

5.2 Radar Network Placement

To monitor the hand marker target in air, three radars have been used to form a radar network. In order to detect the target, the target must be in the FOV

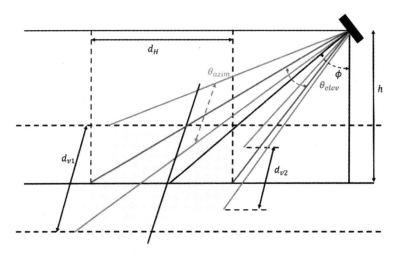

Figure 5.2 FOV of radar placed with a steering angle of ϕ. The gray patch indicates the FOV at height h [16].

of all three radars. Therefore, radar placement plays a vital role in the proposed system. The radar's FOV or coverage depends mainly on the antenna aperture angles θ_{azim}, θ_{elev}. The radar coverage is also related to the target RCS since it determines the received power level.

A radar placed at a height h from the ground as shown in Figure 5.1, then mathematically, the vertical coverage and horizontal coverage is given by

$$\begin{cases} D_{ver} = 2\left[h.tan\left(\frac{\theta_{elev}}{2}\right)\right] \\ D_{hor} = 2\left[h.tan\left(\frac{\theta_{azim}}{2}\right)\right] \end{cases} \tag{5.1}$$

The radar FOV can be extended by steering the radars with an angle ϕ as shown in Figure 5.2 and the FOV (5.1) becomes,

$$\begin{cases} d_H = h\left[\tan\left(\frac{\theta_{elev}}{2} + \phi\right) - \tan\left(\frac{\theta_{elev}}{2} - \phi\right)\right] \\ d_{v1} = 2\left[\frac{h}{\cos\left(\phi + \frac{\theta_{elev}}{2}\right)}\tan\left(\frac{\theta_{azim}}{2}\right)\right] \\ d_{v2} = 2\left[\frac{h}{\cos\left(\phi - \frac{\theta_{elev}}{2}\right)}\tan\left(\frac{\theta_{azim}}{2}\right)\right] \end{cases} \tag{5.2}$$

Keeping in mind the above configurations, the radars were placed to achieve maximum FOV for the virtual drawing board.

5.3 Preprocessing

In FMCW radar, the transmitted signal is frequency modulated by a periodic saw-wave function within a frame. The received wave is subject to time delay τ, where $\tau = \frac{2(R+vt)}{c}$. This time delay manifests itself as beat frequency in the intermediate signal at the receiver after the mixer. Estimates of this time delay are achieved by FFT spectra analysis along each chirp, also denoted as fast time. The three radars in the network are operated noncoherently, that is, the transmit-receive of all radars are active at the same time without time synchronization among them. From the received signal at each radar, the hand marker target localization and tracking in 3D space is achieved by signal processing algorithms outlined in Figures 5.3(a) and (b) and described below.

5.3.1 Coherent Pulse Integration

After computing the FFT along the fast time for each chirp in a frame, pulse integration is applied to combine the data overall chirps to improve the SNR, facilitating subsequent target detection. We here apply coherent integration, which combines the phase and magnitude of the range FFT data coherently over all chirps of a frame. Coherent integration at frame k, $R_{ci}(k)$, is based on the mean of the range FFT data $\{R_i^n(k)\}_{i=1}^{N_C}$ over N_C chirps:

$$R_{ci}^n(k) = \frac{1}{N_C} \cdot \sum_{i=1}^{N_C} R_i^n(k) \tag{5.3}$$

We have used $N_C = 16$, and the number of transmit samples and the length of range bins are 128, 256, respectively.

5.3.2 Moving Target Indication Filtering

In FMCW radar, TX-to-RX leakage is a challenge, which limits the usability of the first few FFT range bins. Further, the reflections from stationary objects can subdue the reflections from the target or hand marker, thus we applied the MTI filter to suppress the contribution of these stationary objects and leakage. Further, in our system, the interference from other radars is manifested in the range FFT spectra, which is also suppressed to some extent by the MTI filtering operation. At each frame, the coherent integrated range FFT spectrum, denoted as $R_{ci}(k)$ at the kth frame, is passed through running average as

$$R_{ci}(k) = R_{ci}(k) - S(k-1) \tag{5.4}$$

$$S(k) = \alpha \cdot R_{ci}(k) + (1-\alpha) \cdot S(k-1) \tag{5.5}$$

where α is the forget factor, set to 0.9. This filtered range FFT spectrum is then utilized for the subsequent target detection.

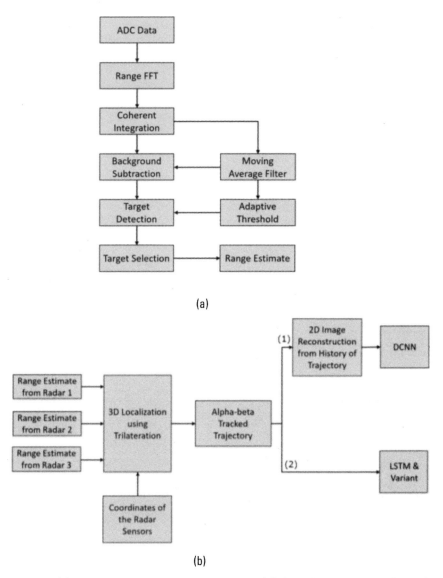

(a)

(b)

Figure 5.3   (a) Range estimate at each radar sensor, and (b) 3D localization through trilateration
and character recognition using (1) DCNN using 2D image reconstruction, and (2)
LSTM using time-series trajectory data [16].

5.3.3 Target Detection and Selection

The target detection is simply done through thresholding with a scaled value of
the mean of the range FFT spectra calculated at every frame. Thus the threshold at
frame k is set as $\beta \sum_{n=1}^{N_s} R_{ci}^n(k)$ where β is the scaling and is set to 3 in our case and
n is the index along range bins. Additionally, the closest detected target is selected

for each radar in the case of multiple detected target peaks. Further, sophisticated data association techniques could also be used for target selection. The interference among FMCW radars are limited by lowering the transmit powers of each radar, and the effects of interference are mitigated by MTI filter; however, further residual interference is taken care of by the appropriate thresholding function.

Further, the range estimates from each radar sensor are combined to localize the 3D space through trilateration, which is described in the following section.

5.3.4 Localization with Trilateration

The localization of the hand marker target is achieved through trilateration, which is the process of determining the global coordinates by utilizing the distance information to some references whose coordinates are known a priori. Among the different approaches and algorithms existing in the literature, we used the approach proposed in [9]. In order to understand the method, let us consider three reference points $R_1(x_1, y_1, z_1)$, $R_2(x_2, y_2, z_2)$, and $R_3(x_3, y_3, z_3)$. Let a target point $T(x, y, z)$ be located at distances D_1, D_2, and D_3, respectively. Then the coordinates of the point T can be obtained by solving the following quadratic equations:

$$\begin{cases} (x - x_1)^2 + (y - y_1)^2 + (z - z_1)^2 = D_1^2 \\ (x - x_2)^2 + (y - y_2)^2 + (z - z_2)^2 = D_2^2 \\ (x - x_3)^2 + (y - y_3)^2 + (z - z_3)^2 = D_3^2 \end{cases} \tag{5.6}$$

Expanding the squares, (5.6) becomes

$$\begin{cases} (x^2 + y^2 + z^2) - 2x_1 x - 2y_1 y - 2z_1 z = D_1^2 \\ \qquad\qquad - x_1^2 - y_1^2 - z_1^2 \\ (x^2 + y^2 + z^2) - 2x_2 x - 2y_2 y - 2z_2 z = D_2^2 \\ \qquad\qquad - x_2^2 - y_2^2 - z_2^2 \\ (x^2 + y^2 + z^2) - 2x_3 x - 2y_3 y - 2z_3 z = D_3^2 \\ \qquad\qquad - x_3^2 - y_3^2 - z_3^2 \end{cases} \tag{5.7}$$

or in matrix form:

$$\begin{bmatrix} 1 & -2x_1 & -2y_1 & -2z_1 \\ 1 & -2x_2 & -2y_2 & -2z_2 \\ 1 & -2x_3 & -2y_3 & -2z_3 \end{bmatrix} \begin{bmatrix} x^2 + y^2 + z^2 \\ x \\ y \\ z \end{bmatrix} = \begin{bmatrix} D_1^2 - x_1^2 - y_1^2 - z_1^2 \\ D_2^2 - x_2^2 - y_2^2 - z_2^2 \\ D_3^2 - x_3^2 - y_3^2 - z_3^2 \end{bmatrix} \tag{5.8}$$

Equation (5.8) can be written in the form:

$$Ax = b \tag{5.9}$$

subject to constraint $x \in C$ where $C = \{(x_0, x_1, x_2, x_3)^T \in \mathfrak{R}^2 / x_0 = x_1^2 + x_2^2 + x_3^2\}$.

The solution to the above problem is provided in [9] and has been used here.

Now to solve the system of equation we have the following two cases.

Case 1: When the references points R_1, R_2, and R_3 do not lie on a straight line, then this makes the *Range*(A) $= 3$ and *dim(Kern*(A)) $= 1$. In such a case, the solution to the system of (5.9) is given by

$$x = x_p + \alpha.x_h \tag{5.10}$$

where x_p is the particular solution of (5.9) and x_h is a solution of the homogeneous system $Ax = 0$ and α is a real parameter. To determine the parameter α, let us suppose $x_p = (x_{p0}, x_{p1}, x_{p2}, x_{p3})^T$, $x_h = (x_{h0}, x_{h1}, x_{h2}, x_{h3})^T$, and $x = (x_0, x_1, x_2, x_3)^T$, and inserting it into (5.10) we get

$$\begin{cases} x_0 = x_{p0} + \alpha.x_{h0} \\ x_1 = x_{p1} + \alpha.x_{h1} \\ x_2 = x_{p2} + \alpha.x_{h2} \\ x_3 = x_{p3} + \alpha.x_{h3} \end{cases} \tag{5.11}$$

By using the constraint $x \in C$ it follows that:

$$x_{p0} + \alpha.x_h0 = (x_{p1} + \alpha.x_{h1})^2 + (x_{p2} + \alpha.x_{h2})^2 + (x_{p3} + \alpha.x_{h3})^2$$

and thus

$$\alpha^2 (x_{h1}^2 + x_{h2}^2 + x_{h3}^2) + \alpha(2.x_{p1}x_{h1} + 2.x_{p2}x_{h2} + 2.x_{p3}x_{h3} - x_{h0}) + x_{p1}^2 + x_{p2}^2 + x_{p3}^2 - x_{p0} = 0 \tag{5.12}$$

Equation (5.12) is a qaudratic equation of the form $a\alpha^2 + b\alpha + c = 0$ with the solutions

$$\alpha_{1/2} = \frac{-b \pm \sqrt{b^2 - 4ac}}{2a}. \tag{5.13}$$

The solutions of (5.9) are

$$x_1 = x_p + \alpha_1.x_h \quad \text{and} \quad x_2 = x_p + \alpha_2.x_h. \tag{5.14}$$

If the distances are too short, then the multilateration porblem cannot be solved and hence there is no real solution [9]. In this case, the real part is used as an approximation for the solution. With this approximation, the constraint $x_{1/2} \in C$ is not met. Thus the difference,

$$d = x_0 - (x_1^2 + x_2^2 + x_3^2) \tag{5.15}$$

is a measure of the solvebility of the multilateration problem, where x_0, x_2, and x_3 are the coordinates of the solution $\mathbf{x}$ in (5.9). Solutions of the multilateration problem are the points:
$T_1 = \mathbf{x}_1.\mathbf{M}$ and $T_2 = \mathbf{x}_2.\mathbf{M}$ where,

$$\mathbf{M} = \begin{bmatrix} 0 & 0 \\ 0 & I \end{bmatrix}, \quad \text{where} \quad I = 3 \times 3 \quad \text{identity.} \tag{5.16}$$

Case 2: When the reference points R_1, R_2, and R_3 lie on a straight line, then the *Range*$(\mathbf{A}) = 2$ and *dim*$(Kern(\mathbf{A})) = 2$. The general solution (5.9) is then

$$\mathbf{x} = \mathbf{x}_p + \alpha.\mathbf{x}_{h1} + \gamma.\mathbf{x}_{h2} \tag{5.17}$$

with real parameter α and γ, where $\mathbf{x}_p$ is a particular solution of (5.9) along with two solutions $\mathbf{x}_{h1}$ and $\mathbf{x}_{h2}$ of the homogenous system $\mathbf{A}\mathbf{x} = \mathbf{0}$. They are linearly independent solutions and form the basis of *Kern*$(\mathbf{A})$. In this case the multiletration problem has infinitely many solutions because there is only one constraint equation.

5.3.5 Trajectory Smoothening Filters

An alpha-beta filter is used to smooth the target trajectory over 3D space. The alpha-beta filter predicts position and velocity assuming zero acceleration of a moving target. It iterates between prediction and update steps and smooths target trajectory. The prediction process is expressed as follows:

$$x(k) = \bar{x}(k-1) + \delta T \cdot v(k-1), \tag{5.18}$$

$$v(k) = \bar{v}(k-1), \tag{5.19}$$

where δT is the measurement update interval or frame time, $x(k)$ and $\bar{x}(k)$ are the predicted and smoothed target position at time $k\delta T$, respectively, and $v(k)$ and $\bar{v}(k)$ are the predicted and smooth target velocities at time $k\delta T$, respectively.
The update process is defined as

$$\bar{x}(k) = x(k) + \alpha(\hat{x}(k) - x(k)), \tag{5.20}$$

$$\bar{v}(k) = v(k) + \frac{\beta}{\delta T}(\hat{x}(k) - x(k)), \tag{5.21}$$

where $\hat{x}(k)$ is the measured position of the target at $k\delta T$ trajectory instant. The values of the correction gains α, β are chosen empirically, and in our case set to

$\alpha = 0.3$ and $\beta = 0.5$. Additionally, whenever a target measurement is outside the defined maximum and minimum 3D coordinates, the update process is skipped.

5.4 Setup and Characters

5.4.1 Character Set

For evaluating the proposed system, we used capital letters A–J and numerals 1–5 for detection and classification. We use the processing pipeline as outlined in Section 5.3. An important topic is detecting the end of a character. For the case of a DCNN-based recognition system, this is achieved by removing the hand marker once a character is written on the virtual board. However, in the case of LSTM-CTC network characters are written continuously on the virtual board and the transient detection between the end of a character and start of a next character is trained as a *blank* character, as is explained later. Thus, a LSTM-CTC-based system does not require an explicit stop of a character in the case of continuous writing.

5.4.2 System Parameters

We use the sweep bandwidth of 6 GHz; hence the theoretical range resolution of our system is $\delta r = c/2B = 2.5$ cm, where c is the speed of light. Due to the fast-time window function, it would result in some loss of range resolution. The number of transmit DAC samples used for generating the chirp is $NTS = 128$; hence the maximum detectable range is $R_{max} = (NTS/2) \cdot \delta r = 1.6$m, since the *BGT60TR24B* sensor has only I channel. The ADC sampling frequency is set to 0.747664 MHz.

The chirp time is set to 171.2 μs and the chirp repetition time is set to 128 μs and we use 16 consecutive chirps in a frame. The frame time is set to 50 ms, and we collect 100 frames (i.e., 10s of data once a moving target/hand-marker is detected). The maximum speed of the hand marker is determined by the frame time, 50 ms, and the minimum speed is determined by the MTI filter, which removes the static targets. Since we use a factor of 0.01, if the speed of the hand marker is less than 5s, the hand marker would be removed (or not detected) as a static target.

We use 1 TX, 1 RX antennas, and the 3-dB azimuth and elevation aperture angles are both 70°. All the system parameters are outlined in Table 5.1.

5.4.3 Setup and Data Acquistion

For our experiments, we used two setups, one with a vertical virtual board and the other with a horizontal virtual board, as presented in Figure 5.4, respectively. The ADC data from the three radars are fed through a USB into a PC. The data recording was automated with the detection of the hand marker or drawing

Table 5.1
Operating Parameters [16]

Parameters, Symbol	Value
Number of radars	3
Range resolution of a single radar	2.5 cm
Elevation θ_{elev} per radar	70°
Azimuth θ_{azim} per radar	70°
Ramp start frequency, f_{min}	57 GHz
Ramp stop frequency, f_{max}	63 GHz
Bandwidth, B	6 GHz
Chirp time, T_c	171.2 μs
Sampling frequency, fs	0.747664 MHz
Number of samples per chirp, Ns	128

pen and terminated once the hand marker was not detected in the frame for 20 consecutive frames. Trajectories for each letter (A–J) and numeral (1–5) were used for our system evaluation. Each drawn character with large interclass variance was recorded and labeled. For each character, 100 samples were recorded for training and 25 samples for testing from random trials.

The change of range of the drawing hand marker as seen from each radar for a few letters are shown in Figure 5.5. First, the distance of the marker from all the radars are calculated and then these distances are fed to the trilateration algorithm. The trilateration algorithm first localizes the marker in the FOV and then tracks the marker motion. In order to smooth the tracking, moving average is used that then feeds the tracked trajectory or reconstructed image to the LSTM/CNN for classification. Examples of the reconstructed 2D image of the unistroke character from the data set are shown in Figure 5.6, which presents the examples that are clean-drawn characters by the user without any extraneous strokes. However, the characters are rotated since in the virtual board there is no reference and users are free to draw based on arbitrary references.

Figure 5.7 presents other examples of reconstructed images of letters and numerals from the written trajectories in the virtual board. These examples illustrate the large variations in the characters drawn, in terms of character size, orientation, and pattern including extraneous strokes within drawn characters. This illustrates the large intraclass variation of the drawn characters, which are affine and nonaffine transforms of the clean characters drawn in Figure 5.6.

In the following sections, we outline two approaches for recognition of drawn characters, (1) LSTM and its variants, and (2) DCNN for recognition of the unknown drawn character as shown in Figure 5.3(b). And finally, we present an alternate approach using 1D CNN-LSTM where much of the preprocessing in terms of trilateration and tracking can be relaxed for character recognition tasks.

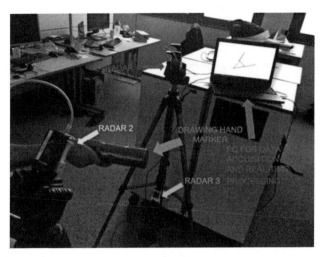

Vertical virtual board setup

(a)

Horizontal virtual board setup

(b)

Figure 5.4   (a) Experimental setup 1 with vertical virtual board, and (b) experimental setup 2 with horizontal virtual board [16].

5.5 LSTM

We cast the problem of classification of the time-series trajectory data as a multiclass classification. Given the temporal trajectory coordinates $\mathbf{x}(k) = (x(k), y(k), z(k))$ at time k, a RNN maps it to a sequence of hidden values $\mathbf{h}(k) = (h_1(k), \cdots, h_T(k))$ and outputs a sequence of activations $\mathbf{a}(k + 1) = (a_1(k + 1), \cdots, a_T(k + 1))$ by iterating the following recursive equation:

$$\mathbf{h}(k) = \sigma\left(W_{hx}\mathbf{x}(k) + \mathbf{h}(k - 1)W_{hh} + \mathbf{b}_h\right) \tag{5.22}$$

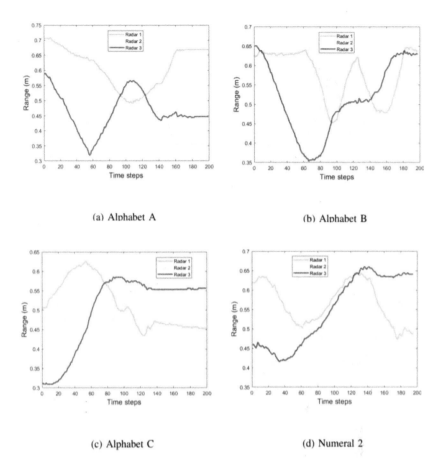

(a) Alphabet A

(b) Alphabet B

(c) Alphabet C

(d) Numeral 2

Figure 5.5 Change of range from the three FMCW radars for letters A, B, C and numeral 2 written on a virtual board [16].

where σ is the nonlinear activation function, $\mathbf{b}_h$ is the hidden bias vector and W terms denote weight matrices, W_{hx} being the input-hidden weight matrix and W_{hh} the hidden-hidden weight matrix. We evaluate the classification performance of the time-series trajectory approach using LSTM, BLSTM, and ConvLSTM with CTC loss function. The activation for these recurrent units is defined by:

$$\mathbf{a}(k+1) = h(k) W_{ha} + \mathbf{b}_a \tag{5.23}$$

where W_{ha} denotes the hidden-activation weight matrix and the $\mathbf{b}_a$ terms denote the activation bias vector.

5.5.1 Architecture

LSTMs extend RNN with memory cells using the concept of gating, a mechanism based on component-wise multiplication of the input, which defines the behavior

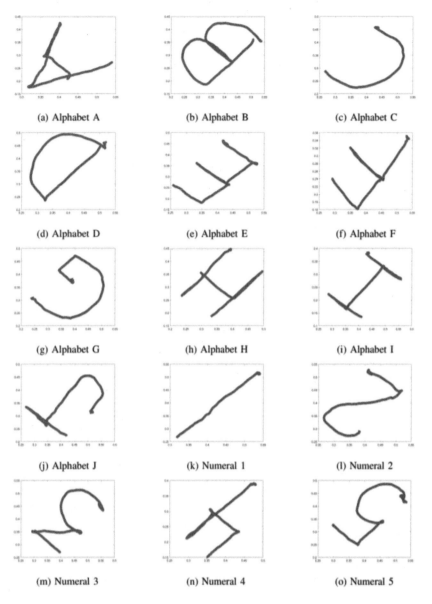

(a) Alphabet A   (b) Alphabet B   (c) Alphabet C

(d) Alphabet D   (e) Alphabet E   (f) Alphabet F

(g) Alphabet G   (h) Alphabet H   (i) Alphabet I

(j) Alphabet J   (k) Numeral 1   (l) Numeral 2

(m) Numeral 3   (n) Numeral 4   (o) Numeral 5

Figure 5.6   2D reconstructed images of letters and numerals from the written trajectories in the virtual board [16].

of each individual memory cell. The LSTM updates its cell state according to the activation of the gates. The input provided to an LSTM is fed into different gates that control which operation is performed on the cell memory: write (input gate), read (output gate), or reset (forget gate). LSTM mitigates the issue of the vanishing gradient in RNNs.

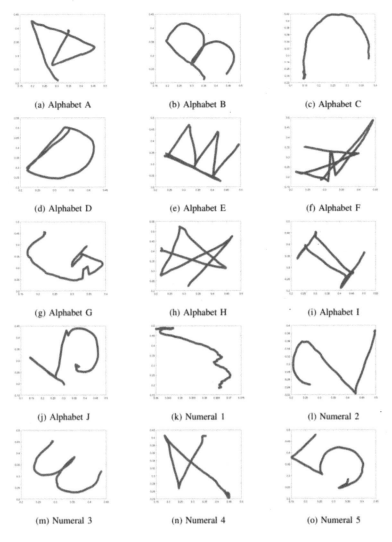

(a) Alphabet A

(b) Alphabet B

(c) Alphabet C

(d) Alphabet D

(e) Alphabet E

(f) Alphabet F

(g) Alphabet G

(h) Alphabet H

(i) Alphabet I

(j) Alphabet J

(k) Numeral 1

(l) Numeral 2

(m) Numeral 3

(n) Numeral 4

(o) Numeral 5

Figure 5.7 Additional examples of 2D reconstructed images of alphabets and numerals from the data set [16].

The vectorial representation (vectors denoting all units in a layer) of the update of an LSTM layer is as follows:

$$
\begin{cases}
\mathbf{i}(k) = \sigma_i(W_{ai}\mathbf{a}(k) + W_{hi}\mathbf{h}(k-1) + W_{ci}\mathbf{c}(k-1) + \mathbf{b}_i) \\
\mathbf{f}(k) = \sigma_f(W_{af}\mathbf{a}(k) + W_{hf}\mathbf{h}(k-1 + W_{cf}\mathbf{c}(k-1) + \mathbf{b}_f) \\
\mathbf{c}(k) = \mathbf{f}(k)\mathbf{c}(k-1) + \mathbf{i}(k)\sigma_c(W_{ac}\mathbf{a}(k) + W_{hc}\mathbf{h}(k-1) + \mathbf{b}_c) \\
\mathbf{o}(k) = \sigma_o(W_{ao}\mathbf{a}(k) + W_{ho}\mathbf{h}(k-1) + W_{co}\mathbf{c}(k) + \mathbf{b}_o) \\
\mathbf{h}(k) = \mathbf{o}(k)\sigma_h(\mathbf{c}(k))
\end{cases} \quad (5.24)
$$

where $\mathbf{i}$, $\mathbf{f}$, $\mathbf{o}$, and $\mathbf{c}$ are, respectively, the input gate, forget gate, output gate, and cell activation vectors, all of which are the same size as vector $\mathbf{h}$ defining the hidden value. Terms σ represent nonlinear functions. The term $\{\mathbf{x}(1), \mathbf{x}(2), \cdots, \mathbf{x}(K)\}$ is the input to the memory cell layer at time k. W_{ai}, W_{hi}, W_{ci}, W_{af}, W_{hf}, W_{cf}, W_{ac}, W_{hc}, W_{ao}, W_{ho}, and W_{co} are weight matrices, with subscripts representing from-to relationships $\mathbf{b}_i$, $\mathbf{b}_f$, $\mathbf{b}_c$, and $\mathbf{b}_o$ are bias vectors.

BLSTMs [14], is a variant of LSTM, which duplicates the LSTM layer such that one LSTM layer takes input of the input sequence, the same as in (unidirectional) LSTM, and another takes the reversed copy of the input sequence to the other LSTM layer. BLSTMs capture temporal information of the trajectory in the time-series data as the characters are drawn in air. Since the coordinates are run both forward and backward, BLSTM is able to preserve information from the future and then concatenates the two hidden states for prediction.

ConvLSTM [15] is another variant of LSTM, which instead of fully connected operations in input-to-state and state-to-state transitions, uses convolution. ConvLSTM, unlike LSTM & BLSTM, preserves the spatial information of the trajectory coordinates instead of flattening the input vector. ConvLSTM replaces matrix multiplication with convolution operation at each gate in the LSTM cell; that is, in the case of $\mathbf{i}(k)$, $\mathbf{f}(k)$, $\mathbf{o}(k)$ in (5.22). ConvLSTM has been shown to be quite effective in multidimensional time-series data on several problems and is also evaluated for this problem.

The multiclass classification using LSTM, BLSTM, and ConvLSTM with the class output is depicted in Figure 5.8. The input sequence $\mathbf{x}(1 : t), \mathbf{x}(2 : t + 1), \cdots, \mathbf{x}(K - t : K)$ indicates the input to the LSTM layer at each time step k with t denoting the window length of the trajectory coordinates. Thus the input size to LSTM is $3t$, where t is used as hyperparameter and 3 denotes the target coordinates. LSTM and its variants have a hidden state size of 150, which is further fed to a dense connection of size 15 atop followed by a sigmoid activation function. As described earlier, for BLSTM the two LSTM connections are concatenated before feeding to the dense layer.

5.5.2 Loss Function: CTC

One of the biggest challenges in sequence problems is addressing the issue of variable timing. An effective practice to tackle such problems is the use of CTC [10,11] operation that removes the need of having an aligned data set and makes the training much easier. In our problem statement, we use CTC operation to enable detection of *blank* class, which marks the end of a character or start of a new character being drawn on the virtual space and the marker is in transit from one position to another without marked trajectory.

A new class *blank* class is appended to the existing dictionary containing trajectories of supported characters such that

$$\mathbf{C}' = \{C_1, C_2, \cdots C_{15}\} \cup \{blank\} \tag{5.25}$$

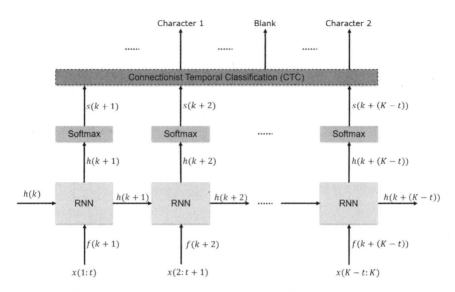

Figure 5.8 RNN networks; that is, LSTM, BLSTM, ConvLSTM architectures, with CTC loss function for prediction of unsegmented character recognition, depicting the input with t features at each time step with a window stride of 1 along the temporal dimension [16].

The class-conditional probability of the class is given by the application of softmax. The probability of occurrence of a particular m at time k in an input sequence is computed by the network and can be represented as:

$$\mathsf{X} : P(m, k|\mathsf{X}) = s_k^m \ \forall k \in [0, K). \tag{5.26}$$

An input sequence X can be mapped into a sequence of class labels $\mathbf{y}$ by a defined path π and the probability of such observance is defined as

$$P(\pi, \mathsf{X}) = \prod_k s_k^{\pi_k} \tag{5.27}$$

where π_k is the predicted class at time k in path π.

A single-label sequence may correspond to multiple paths that can yield repetition of labels. The condensing of repeated labels as well as removal of blank labels is performed by employing a many-to-one function β. For example, $\beta([-, A, B, -, -]) = \beta([A, -, B, -]) = [A, B]$ where A and B are true labels and $'-'$ is *blank*.

The sum of all conditional probabilities of different paths π mapping to the given input sequence X gives the probability of observing a particular $\mathbf{y}$. This can be represented as $\beta^{-1}(\mathbf{y}) = \{\pi : \beta(\pi) = \mathbf{y}\}$,

$$P(\mathbf{y}|\mathsf{X}) = \sum_{\pi \in \beta^{-1}(\mathbf{y})} P(\pi|\mathsf{X}) \tag{5.28}$$

The probability $P(y|X)$ can be computed by dynamic programming [10,11]. The final CTC loss function can be written as:

$$L_{CTC} = -\ln P(y|X) \tag{5.29}$$

5.5.3 Design Considerations

To avoid overfitting we have used recurrent dropout without memory loss [12], where the dropout is only applied to the incremental memory cell update.

Typically, LSTM and its variants' parameters are initialized to small random values in an interval; for example, in our experiments, we initialize weights and biases to random values in the interval [-0.1, 0.1]. However, for the forget gate, this is a suboptimal choice, where small weights effectively close the gate, preventing cell memory and its gradients from flowing in time. One approach to address this issue is to initialize the forget gate bias, b_f, to a large value, say 1, that will force the gate to be initialized in an open position and allow memory cell gradients in time to flow more readily. We trained the LSTM network and its variants for 100 epochs with early stopping criteria.

5.5.4 Performance Evaluation

The results of the classification were evaluated by an accuracy measure. Table 5.2 compares the performance of different LSTM variants using CTC loss function in terms of model size and accuracy using the window length t as hyperparameter. As outlined in Table 5.2, ConvLSTM-CTC performs the best in terms of accuracy on the time-series trajectory data. This is explainable since ConvLSTM preserves the spatiotemporal structure of the input trajectory data, unlike LSTM and BLSTM. Additionally, as observed from the Table 5.2 the model size remains the same for LSTM and BLSTM as each input time step is taken with a stride of 1, whereas the size decreases with increasing t as the stride is also set to t.

Table 5.2

Comparison of Model Accuracy and Size for Different LSTM
Variants with Varying Input Features [16]

LSTM Variants	t = 1		t = 5		t = 10	
	Acc.	Size	Acc.	Size	Acc.	Size
LSTM-CTC	93.33	1.10	93.33	1.15	91.67	1.22
BLSTM-CTC	78.33	8.38	95	8.50	96.67	8.64
ConvLSTM-CTC	93.33	3.54	95	1.78	98.33	1.56

$t = 1, 5, 10$, accuracy is in % and size in MB.

Table 5.3
Confusion Matrix of the Proposed ConvLSTM-CTC [16]

		Predicted Class													
		A	**B**	**C**	**D**	**E**	**F**	**G**	**H**	**I**	**J**	**1**	**2**	**3**	**4**	**5**
	A	95	0	0	0	0	0	0	5	0	0	0	0	0	0	0
	B	0	100	0	0	0	0	0	0	0	0	0	0	0	0	0
	C	0	0	100	0	0	0	0	0	0	0	0	0	0	0	0
	D	0	0	0	100	0	0	0	0	0	0	0	0	0	0	0
	E	0	0	0	0	95	0	0	0	0	0	0	0	5	0	0
Actual Class	**F**	0	0	0	0	0	95	0	0	0	0	0	0	0	0	5
	G	0	0	0	0	0	0	100	0	0	0	0	0	0	0	0
	H	0	0	0	0	0	0	0	100	0	0	0	0	0	0	0
	I	0	0	0	0	0	0	0	0	95	5	0	0	0	0	0
	J	0	0	0	0	0	0	0	0	0	100	0	0	0	0	0
	1	0	0	0	0	0	0	0	0	0	0	100	0	0	0	0
	2	0	0	0	0	0	0	0	0	0	0	0	100	0	0	0
	3	0	0	0	0	0	0	0	0	0	0	0	0	100	0	0
	4	0	0	0	0	0	0	0	0	0	0	0	0	0	100	0
	5	0	0	0	0	0	0	0	0	0	0	0	0	0	0	100

Input features $t = 10$ on test set containing 25 examples from all classes. The overall accuracy is 98.33%.

Table 5.3 show the confusion matrices obtained for testing ConvLSTM with the aforementioned testing data set [16]. ConvLSTM achieves a classification accuracy of 98.33%. As observed from Table 5.3, characters with similar trajectories are confused by ConvLSTM on the test data set, such as A with H, I with J, E with 3, and F with 5. ConvLSTM-CTC supports variable-length characters, thus real-time continuous writing of characters in a word, thus paving the way for a more sophisticated system.

5.6 Deep Convolutional Neural Networks

For the case of character recognition using DCNN, we first reconstruct the trajectory of the hand marker over the 3D space in a 2D plane by stitching together the coordinates from the start to the end of the drawn character.

5.6.1 Architecture

The DCNN takes an input of 64×64 binary image that has 6 convolution layers. Each convolution layer is followed by a batch normalization layer and max-pooling layer of size 2×2. The first convolution layer uses 8 convolutional filters with a size of 3×3. The second, third, fourth, fifth, and sixth convolutional

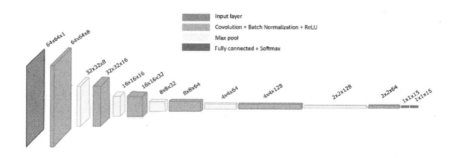

Figure 5.9 Proposed DCNN architecture for recognition of a character from the 2D reconstructed image from the hand marker trajectory [16].

layers use 16, 32, 64, 128, and 64 convolutional filters with a size of 3×3, respectively. All of the convolutional layers use ReLUs as an activation function and batch normalization is applied prior to ReLU. Following the convolutional layers, a fully connected layer with a softmax activation function is appended. The proposed DCNN architecture is depicted in Figure 5.9.

5.6.2 Weight Initialization

The weight initialization for 2D convolutional layers was performed by drawing random samples from a uniform distribution between $(-limit, limit)$, where $limit = \sqrt{\frac{6}{n_i + n_o}}$ and n_i and n_o are the number of input and output units, respectively. The weights of the fully connected hidden layers were initialized by drawing random samples from a normal distribution N (0, 0.01) and the biases were initialized with samples drawn from a normal distribution but with mean 0.5.

5.6.3 Learning Schedule

We use an Adam [25] optimizer that computes the adaptive learning rate for each network weight over the learning process from estimates of the first and second moments of the gradients. In configuration parameters, the learning rate (alpha) is set to 0.001 and the exponential decay rate for the first (beta1) and second (beta2) moment estimates are set to 0.9 and 0.999. The epsilon that counters divide by zero problems is set to 1e-8.

5.6.4 Data Augmentation

Data augmenting with transformed versions of the training samples helped improve the generalization capability of the DCNN model, particularly when we used the following transformations:

- Random rotation, which rotates the 2D image along a random axis changing the reference about which the character was drawn;
- Random scaling, which scales the size of the 2D image to account for users drawing varying character sizes;
- Random skew, which skews the training character data to account for variabilities in the user's drawn character;
- Random translation, which translates the data set with arbitrary start and end coordinates.

5.6.5 Performance Evaluation

Table 5.4 shows the confusion matrices obtained for testing DCNN with the aforementioned testing data set. DCNN achieves a classification accuracy of 98.33%, and the recognition system confuses numeral 4 with letter J on the reconstructed images.

In the case of DCNN, we found that a number of procedures helped to increase the accuracy of the system. Weight decay and dropout prevented the network from overfitting to the training data and improved the classification accuracy by 4% to 5% on average. Augmenting the training data set with

Table 5.4

Confusion Matrix of the Proposed DCNN on Test Set Containing 25 Examples from All Classes [16]

		Predicted Class													
		A	B	C	D	E	F	G	H	I	J	1	2	3	4	5
Actual Class	A	100	0	0	0	0	0	0	0	0	0	0	0	0	0	0
	B	0	100	0	0	0	0	0	0	0	0	0	0	0	0	0
	C	0	0	100	0	0	0	0	0	0	0	0	0	0	0	0
	D	0	0	0	100	0	0	0	0	0	0	0	0	0	0	0
	E	0	0	0	0	100	0	0	0	0	0	0	0	0	0	0
	F	0	0	0	0	0	100	0	0	0	0	0	0	0	0	0
	G	0	0	0	0	0	0	100	0	0	0	0	0	0	0	0
	H	0	0	0	0	0	0	0	100	0	0	0	0	0	0	0
	I	0	0	0	0	0	0	0	0	100	0	0	0	0	0	0
	J	0	0	0	0	0	0	0	0	0	100	0	0	0	0	0
	1	0	0	0	0	0	0	0	0	0	0	100	0	0	0	0
	2	0	0	0	0	0	0	0	0	0	0	0	100	0	0	0
	3	0	0	0	0	0	0	0	0	0	0	0	0	100	0	0
	4	0	0	0	0	0	0	0	0	0	20	0	0	0	80	0
	5	0	0	0	0	0	0	0	0	0	0	0	0	0	0	100

The overall accuracy is 98.33%.

transformed versions of the training samples also helped to improve the generalization capability of the DCNN. The affine transformations particularly helped in improving classification performance in the case of characters differing by one or two strokes, such as I and J.

The proposed architectures are quite simple but they still achieve very robust recognition performance on letters and numerals in real time.

5.7 1D CNN-LSTM

The motion trajectory forms a letter that resembles unistroke writing, and is the foundation for recognition of complex words and the focus of our future work. Furthermore, in this work, we have used a hand marker for the drawn characters, as in future work we aim to replace a hand marker with only a hand. Hand-only poses challenges due to multitarget reflections from other body parts at the three radar sensors and occlusion of the hand at one or more radar sensors at multiple instances. The trilateration and single-target tracking/smoothening algorithms presented in an Section 5.3 would fail in these scenarios.

The idea of 1D CNN in conjunction with LSTM is to enable a mechanism to make the solution work if detection from one of the radar at a particular time is not available. When this happens, the earlier approach using trilateration and a tracking filter would fail. However we demonstrate here that if 1D CNN is used as a feature extraction and preprocessing instead, even if one or two of the channels, such as radar sensor data, are nulled, the feature extractor followed by LSTM is able to recognize the characters. The solution is depicted in Figure 5.10.

5.7.1 Architecture

The architecture of the 1D CNN is described in Figure 5.11, wherein the input range estimates, in the form of range spectrum, from the three radar sensors are fed in as seperate channels to the 1D CNN. The 1D CNN progressively applies 1D convolutional operation followed by max-pooling to reduce the spatial dimension and increase the global reception field.

5.7.2 Performance Evaluation

Table 5.5 presents the performance of the proposed 1D CNN-LSTM architecture which range spectra from three radars, two radars, and one radar are used as input. When range spectrum data from all three radars are fed into the network, the recognition accuracy is 98.33%, and when range spectrum data from two radars are fed the accuracy of the network is still intact at 98.33%. This result is remarkable since in the earlier approaches the preprocessing would fail completely, leading to no recognition, which would be the case for finger movement in the

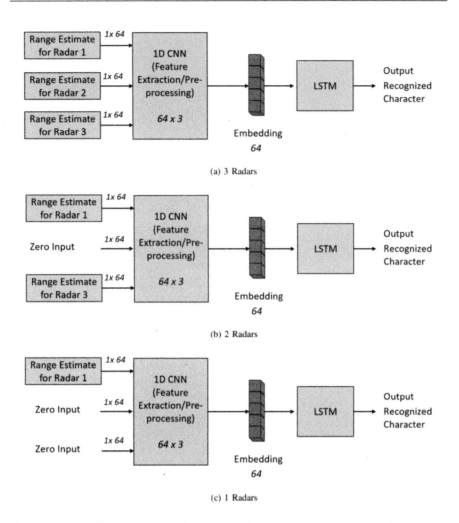

Figure 5.10   1D CNN-LSTM using (a) 3, (b) 2, and (c) 1 valid range spectrum noncoherent radar sensors.

air. When range spectrum data from only one radar sensor is fed, the recognition accuracy falls to 93.33% due to the absence of data. The 1D CNN-LSTM solution offers a promising result for recognition of hand movements on an imaginary backboard drawn in air.

5.8   Future Work and Directions

In the presented solutions, hand markers have been used for character drawing in the imaginary board generated by the radar networks. For the 2D CNN and LSTM based approach, the solution fails if one of the radars doesn't have detection

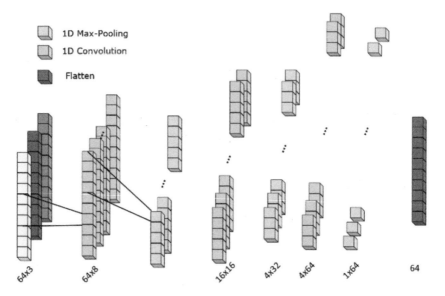

Figure 5.11 Proposed 1D-CNN architecture for preprocessing to act as a feature extraction instead of a trilateration and alpha-beta tracker for smoothing the trajectory.

Table 5.5
Comparison of Model Accuracy and Size
for Different LSTM Variants with Varying
Input Features

Valid Range Spectrum	Accuracy (in %)
All three radars	98.33
Two radars	98.33
One radar	93.33

$t = 1, 5, 10$, accuracy is in % and size in MB.

for a long enough time due to occlusion or shadowing since the trilateration algorithm relies on all three radars' data. Although the 1D CNN-LSTM based solution is able to alleviate this problem by a small performance drop, more neural network architecture search is required to make sure the character recognition system is capable of maintaining the same accuracy under a sparse radar network, such as a two- or one-radar sensor. This would enable the system to operate even with human fingers drawing the characters on a virtual blackboard.

Furthermore, such a system placed on laptop or desktop monitors, such as the bassel corner of a monitor would exhibit blind spots due to the radar's field of view. Solutions to deal with blind spots need to be solved for the solution to be adopted in monitors and AR-VR screens.

5.9 Problems

1. State the equation defining the vertical field of view and horizontal field of view for radar with θ_v and θ_h opening angle placed at a sheering angle of ϕ and height h from ground.
2. What are the different approaches presented in the chapter to combine the data from multiple radars in the network to a common reference? Is there any adverse impact of noncoherent data combining and if so, why?
3. Propose the processing split for central and distributed processing; that is, which operations should be performed locally and which should be performed centrally in a practical efficient deployment.
4. State the difference between ConvLSTM from LSTM. When is ConvLSTM better than LSTM?
5. What is the need for a tracking algorithm to smooth the data before feeding to DCNN or LSTM? How does 1D CNN-LSTM implementation handle this operation?
6. Define CTC loss. What is the advantage of using CTC loss compared to cross entropy for training LSTM in the context of air-writing.
7. Define the processing pipeline if DCNN has to be replaced by a conventional machine learning algorithm. [Hint: How to handle arbitrary reference.]

References

[1] Fan, T., C. Ma, Z. Gu, et al., "Wireless Hand Gesture Recognition Based on Continuous-Wave Doppler Radar Sensors," *IEEE Transactions on Microwave Theory and Techniques*, Vol. 64, No. 11, 2016, pp. 4012–4020.

[2] Molchanov, P., X. Yang, S. Gupta, K. Kim, S. Tyree, and J. Kautz, "Online Detection and Classification of Dynamic Hand Gestures with Recurrent 3d Convolutional Neural Network," in *Proceedings of the IEEE Conference on Computer Vision and Pattern Recognition*, 2016, pp. 4207–4215.

[3] Molchanov, P., S. Gupta, K. Kim, and K. Pulli, "Multi-Sensor System for Drivers Hand-Gesture Recognition," in *Proc. IEEE 11th Int. Conf. Workshops Automat. Face Gesture Recognit.*, Ljubljana, Slovenia, 2015, pp. 1–8.

[4] Moazen, D., S. A. Sajjadi, and A. Nahapetian, "AirDraw: Leveraging Smart Watch Motion Sensors for Mobile Human Computer Interactions," 2016 13th IEEE Annual Consumer Communications and Networking Conference (CCNC 2016).

[5] Amma, C., M. Georgi, and T. Schultz, "Airwriting: Hands-Free Mobile Text Input by Spotting and Continuous Recognition of 3D-Space Handwriting with Inertial Sensors," International Symposiumon Wearable Computers (ISWC), 2012.

[6] Agarwal, C., D. P. Dogra, R. Saini, and P. P. Roy, "Segmentation and Recognition of Text Written in 3D Using Leap Motion Interface," *Proceedings of the 3rd IAPR Asian Conference on Pattern Recognition (ACPR 2015).*

[7] Zhang, X., Z. Ye, and L. Jin, and Z. Feng, and S. Xu, "A New Writing Experience: Finger Writing in the Air Using a Kinect Sensor," *IEEE Multimedia*, Vol. 20, No. 4, 2013, pp. 85–93.

[8] Roy, P., S. Ghosh, and U. Pal, "A CNN Based Framework for Unistroke Numeral Recognition in Air-Writing," 2018 16th International Conference on Frontiers in Handwriting Recognition (ICFHR).

[9] Norrdine, A., "An Algebraic Solution to the Multilateration Problem," 2012 International Conference on Indoor Positioning and Indoor Navigation, Sydney, Australia, 2012.

[10] Molchanov, P., X. Yang, S. Gupta, K. Kim, S. Tyree, and J. Kautz, "Online Detection and Classification of Dynamic Hand Gestures with Recurrent 3D Convolutional Neural Networks," 2016 IEEE Conference on Computer Vision and Pattern Recognition (CVPR), Las Vegas, NV, 2016, pp. 4207–4215.

[11] Zhang, Z., Z. Tian, and M. Zhou, "Latern: Dynamic Continuous Hand Gesture Recognition Using FMCW Radar Sensor," *IEEE Sensors Journal*, Vol. 18, No. 8, April15, 2018, pp. 3278–3289.

[12] Semeniuta, S., A. Severyn, and E. Barth, "Recurrent Dropout without Memory Loss," 26th International Conference on Computational Linguistics, Proceedings of the Conference: Technical Papers, December 11–16, 2016, Osaka, Japan, pp. 17571766.

[13] Kingma, D. P., and J. Ba, "Adam: A Method for Stochastic Optimization," arXiv preprint arXiv:1412.6980, December, 22, 2014.

[14] Xuezhe, M., and E. Hovy, "End-to-End Sequence Labeling via Bi-Directional LSTM-CNNs-CRF," arXiv preprint arXiv:1603.01354, 2016.

[15] Xingjian, S. H. I., Z. Chen, H. Wang, D.-Y. Yeung, W.-K. Wong, and W.-C. Woo, "Convolutional LSTM Network: A Machine Learning Approach for Precipitation Nowcasting," *Advances in Neural Information Processing Systems*, 2015, pp. 802–810.

[16] Arsalan, M., and A. Santra, "Character Recognition in Air-Writing Based on Network of Radars for Human-Machine Interface," *IEEE Sensors Journal*, Vol. 19, No. 19, Oct. 1, 2019, pp. 8855–8864.

6

Material Classification

6.1 Introduction

Material classification has immense potential for industrial, automotive, and consumer applications. In the case of autonomous or self-driving cars, the system must understand the road surface (i.e., tar, pavement, snows or water after a rainfall), it is driving on so that it can adaptively change its mode best suited for the environment. Even industrial and consumer robots require material classification abilities to understand their surroundings for navigation in any environment and also grab objects as per their operational requirements. As well, room-cleaning robots need reliable material classification systems to classify the material it is moping and change their cleaning mode based on a carpet surface, tile, marble, or laminate surface. Such cleaning robots also need to detect liquid on the mopping surface to avoid damage to their systems. Further, material classification is important in packaging and industrial settings, where material identification is required through the packages without damaging the packed goods and luggage [1] or drilling of walls, where pipes or cables underneath are required to be identified for safe drilling.

In [2] a tactile and thermal-based material classification approach with recursive Bayesian estimation is presented. In [3], a material classification machine learning approaches based on a tactile-sensing chip is presented. However, such sensors require physical contact to the material, which in most applications are not possible, apart from high cost. The classification of materials is also a long-standing research topic [4,5] in the field of computer vision. But both these approaches seem to fail if the capture conditions do not match to the training conditions, for example through poor lighting, bad viewing angle, or occlusions.

Recent developments make radar more viable because recent innovations continue to decrease the size, lower power consumption, and lower costs [6]. Through rapid progress in silicon and packing technologies, radar is miniaturized and less expensive. In comparison to other sensors, short-range radar systems have the advantage of being able to operate remotely without requiring to touch the material to be identified, and thus it can sense through the luggage bag or packages to identify material underneath the luggage. Further, radar sensors compared to vision-based systems or infrared-based systems are independent of the operating illumination conditions. This can be immensely advantageous for automotive and industrial use cases, where the working conditions can be quite harsh. Short-range radars are attractive for low-cost handheld consumer devices owing to low maintenance costs and are well suited for material scanning, and through-the-wall scanning for drilling purposes.

In [7], the authors propose a ultrasonic sensor in an automotive setting to classify asphalt, mastic asphalt, grass, gravel, or dirt road in on-road and off-road driving conditions. In [8], the authors use backscattered radar and ultrasonic reflections for road surface recognition in on-road and off-road driving conditions. In [9], the authors present a novel ultrasound-based nondestructive testing system on ceramic using a fusion of a genetic algorithm and artificial neural network. In [10], a material classification system using a time-of-flight (ToF) camera is presented, and a key finding of the study has been that the depth measurement for some materials are distorted as a function of modulating frequency, material type, and distance to the sensor. The authors use these variations to develop a ToF-based material classifier.

6.1.1 Related Work

Automotive radars have been used for classification of road type and discrimination between pedestrian and different vehicles [11]. Ground-penetrating radars are used to detect and identify buried objects such as pipes or mines [12]. Radars have been used to detect materials through luggage or clothes thus enabling material detection for security purposes. In [13], the authors propose a neural network for automotive radar road classification. In [14], the authors present quality control of aeronautics composite multilayered materials and structures using nondestructive testing mechanisms using FMCW radar at 100-GHz and 300-GHz carrier frequency.

Short-range radar sensors, such as RadarCat, use raw ADC data along their four receive antennas in conjunction with random forest to accurately classify everyday objects, material, and body parts [15]. In [16], the authors present a material identification system using a three-dimensional radiance map of the material using a convolutional neural network and demonstrate that processing radiance volume is more efficient than raw signals. In [17], the authors present an AquaCat system that uses radar signals to classify among different liquid types. In [18], the authors present a radar-based system for counting, ordering, and identification of objects such as playing cards, sheets of paper, and Legoblocks.

There are several challenges toward practical deployment of radar-based material identification systems. One challenge is the sensitivity of the system to target angle, sensor orientation, and noise variations due to temperature changes and background. A second challenge is that to sense the surface material,such short-range radar systems will have only a few measurement points along the range dimension, which is a problem for a conventional convolutional neural network with convolutional layers and max-pooling since the dimensions along this dimension would shrink to a very few, losing important information necessary to make a reliable classification. A third challenge is that the sensors produced from different wafer lots have varying extents of manufacturing artifacts and nonlinear response, and thus a classification model trained on one group of sensors does not generalize to other groups from different lots with different characteristics. The last challenge is the problem of material variability; for example, a classifier to distinguish between carpet and tile could be trained with only a certain type that can be made available during the training phase. However, during the life cycle of the system the model could be presented with a carpet or tile type that was not part of the training data set. Further, the classifier should also be capable of rejecting an unknown material as unknown. Training the sensor system using a conventional deep learning algorithm while accounting for all such variations of the data set is practically infeasible.

This chapter is laid out as follows. In Section 6.2 we present the preprocessing and generation of feature images, namely range angle images, that are important for classification of materials. A DCNN-based solution using dilated layers and residual layers is presented in Section 6.3. In Section 6.4, we present the challenges associated with a material classification system in a practical deployment, which we propose to address through a Siamese network. We conclude in Section 6.5 with future research directions and work.

6.2 Features: Range Angle Images

The issue of model sensitivity to target aspect angle, sensor orientation, and noise variations can be mitigated to good extent through robust input feature images. Input feature images that only cause affine transformations due to target aspect angle and sensor orientation and are able to suppress the effects of noise to a good extent are sought after. One such feature image is an RAI generated through a Capon or maximum variance distortionless response (MVDR) algorithm. It can be shown that such RAIs are much more robust compared to the ADC data

Figure 6.1 Example materials are (a) book, (b) chocolate cream, (c) glass, (d) aluminum surface of a laptop, (e) wood, (f) wool, (g) carpet, (h) tile, (i) laminate, and (j) water [20].

or the range spectrum, which are extremely sensitive to aspect angle and noise perturbations.

Figure 6.1 depicts ten everyday materials and objects to demonstrate the material classification system.

Radar imaging is a signal processing technique to map the reflective distribution of the target from multifrequency, multiaspect data within the transmit beam. For a MIMO radar with $N_t = 2$ transmit (Tx) and $N_r = 2$ receive elements (Rx), there are $N_t \times N_r = 4$ distinct propagation channels from the Tx array to the Rx array in a linear array configuration for azimuth angle profiling. If the transmitting source (Tx channel) of the received signals can be identified at the Rx array, a virtual phased array of $N_t \times N_r$ elements can be synthesized with $N_r + N_t$ antenna elements. A time-division multiplexed MIMO array provides a low-cost solution to a fully populated antenna aperture capable of near-field imaging [19].

Denoting the 3D positional coordinates of the Tx element as d_m^{Tx}, $m = 1, 2$ and the Rx element as d_n^{Rx}, $n = 1, 2$ in space, then on assuming far-field conditions, the signal propagation from a Tx element d_m^{Tx} to a point scatterer p and subsequently the reflection from p to Rx element d_n^{Rx} can be approximated as $2x + d_{mn} \sin(\theta)$. Where x is the base distance of the scatterer to the center of the virtual linear array, d_{mn} refers to the position of the virtual element to the center of the array, and θ is the incident angle to the center of the array with respect to boresight. Using time-switching transmission from both transmit antennas, an orthogonal signaling is achieved and at the receiver the virtual antenna array placements is provided by $d^{Tx} * d^{Rx}$ with $*$ representing the spatial convolution of the Tx and Rx antenna locations.

Assuming that far-field conditions are satisfied, the time delay of the radar return from a scatterer at base distance x from the center of the virtual linear array can be expressed as

$$\tau_{mn} = \frac{2x}{c} + \frac{d_{mn} \sin(\theta)}{c} \tag{6.1}$$

The transmit steering vector is expressed as

$$a_m^{Tx}(\theta) = \exp\left(-j2\pi \frac{d_m^{Tx} \sin(\theta)}{\lambda}\right); \quad m = 1, 2 \tag{6.2}$$

while the receiving steering vector is

$$a_n^{Rx}(\theta) = \exp\left(-j2\pi \frac{d_n^{Rx} \sin(\theta)}{\lambda}\right); \quad n = 1, 2 \tag{6.3}$$

where λ is the wavelength of the transmit signal. The $N_t \times N_r$ deramped beat signal can be stacked into a vector and the Kronecker product of the steering vector of the Tx array $a^{Tx}(\theta)$ and the steering vector of the Rx array $a^{Rx}(\theta)$; that is, $a^{Tx}(\theta) \otimes a^{Rx}(\theta)$ can be used to resolve the relative angle θ of the scatterer. Subsequently, beamforming of the MIMO array signals can be regarded as synthesizing the received signals with the Tx and Rx steering vectors.

The azimuth imaging profile for a range bin l can be generated using the Capon spectrum from the beamformer. The Capon beamformer is computed by minimizing the variance/power of noise while maintaining a distortionless response toward a desired angle. The corresponding quadratic optimization problem is

$$\min_{w} w^H C w \text{ s.t. } w^H \left(a^{Tx}(\theta) \otimes a^{Rx}(\theta)\right) = 1 \tag{6.4}$$

Where C is the covariance matrix of noise, the above optimization has a closed-form expression given as $w_{\text{capon}} = \frac{C^{-1} a(\theta)}{a^H(\theta) C^{-1} a(\theta)}$, with being θ a desired angle. On substituting w_{capon} in the objective function of (6.4), the spatial spectrum is given as

$$P_l(\theta) = \frac{1}{\left(a^{\mathrm{Tx}}(\theta) \otimes a^{\mathrm{Rx}}(\theta)\right)^H C_l^{-1} \left(a^{\mathrm{Tx}}(\theta) \otimes a^{\mathrm{Rx}}(\theta)\right)} \quad \text{with } l = 0, ..., L$$

However, estimation of noise covariance at each range bin l is difficult in practice, hence $\hat{C}_l$ is estimated, which contains the signal component as well and can be estimated using the sample matrix inversion (SMI) technique $\hat{C}_l = \frac{1}{N} \sum_{k=1}^{K} s_l^{\mathrm{IF}}(k) s_l^{\mathrm{IF}}(k)^H$ where K denotes the number of snapshots used for signal-plus-noise covariance estimation and $s_l^{\mathrm{IF}}(k)$ is the deramped intermediate frequency signal at range bin l with k being the frame index [20].

Capon spatial spectrum can be viewed as an adaptive signal-dependent spatial filtered spectrum that is characteristics or defined by the material the radar is illuminating. Capon spectra capture the spatial contour of the material intensity reflection and thus are more robust to increase in noise power, antenna array errors, and target aspect angle.

A 60-GHz short-range radar operating from 57 GHz to 64 GHz having 7-GHz bandwidth, thus resulting in a range resolution of 2.14 cm, is used for the Capon images of the objects. The number of ADC samples are taken to be 128 with a chirp time of 32 μs with a sampling rate of 2 MHz. Two transmit and two recieve antennas are operated in MIMO configuration to create an azimuth sweep for each range dimension. The covariance matrix is estimated using the SMI technique using data from 26 consecutive frames. Figure 6.2 presents examples of through Capon-based MIMO imaging principles generated range angle features images $P_l(\theta)$. We use $L = 7$ and $\theta \in [-30°; 30°]$.

6.3 Deep Convolutional Neural Networks

Since the application requires sensing only the surface of the material, the feature image has only a few data points in the range dimension, as shown in Figure 6.2, where $L = 8$, the standard convolution operation of convolution followed by MaxPooling operation would result in too much loss of information in that dimension. It must be noted that the MaxPooling layer is an essential operation for ensuring that the later neurons in the network have a broader global receptive field of the input image. In such cases, the neurons at later layers would learn to base decisions only on local artifacts of the image and thus the network won't generalize. To circumvent the problem of DCNN processing of images with small dimension, dilated convolution can be used instead of the conventional convolution.

Figure 6.3 depicts how a dilated convolution with rates of 1, 2, and 3 work where dilation rate of 1 is the same as the conventional convolution.

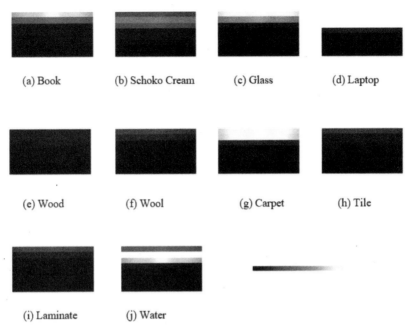

Figure 6.2   Example of range angle feature images of (a) book, (b) chocolate cream, (c) glass, (d) aluminum surface of a laptop, (e) wood, (f) wool, (g) carpet, (h) tile, (i) laminate, and (j) water. In this case, the radar sensed each material roughly 2 cm above the material. The x-axis represents the angle axis and y-axis represents the range axis [20].

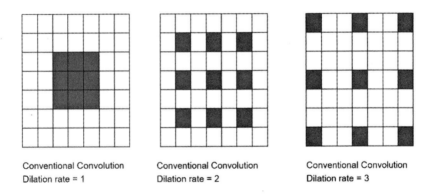

Figure 6.3   Illustration of dilated convolution with rates of 1, 2, and 3.

Compared to standard convolution operation, $\sum_{n+m=k} f(n)g(m)$ where $f(.)$ and $g(.)$ are the input images and the kernel, respectively, dilated convolution can be expressed as $\sum_{n+rm=k} f(n)g(m)$ where r represents the dilation rate. As described, for $r = 1$, we are back to the conventional convolution. Figure 6.4(a-c)

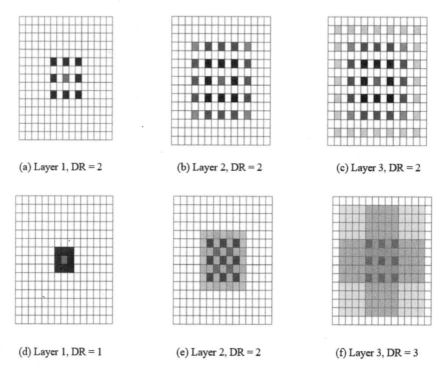

(a) Layer 1, DR = 2   (b) Layer 2, DR = 2   (c) Layer 3, DR = 2

(d) Layer 1, DR = 1   (e) Layer 2, DR = 2   (f) Layer 3, DR = 3

Figure 6.4   Gridding effect on a global receptive field due to dilated convolution with the same dilation rate on: (a) layer 1, (b) layer 2, (c) layer 3 compared to hierarchial dilation of dilation rates of 1, 2, and 3 on (d) layer 1, (e) layer 2, and (f) layer 3.

presents the problem of a gridding effect on the receptive field of the input image due to application of a dilated convolutional of rate 2 across layers 1, 2, and 3. Successive dilated convolution leads to holes in the receptive field of the input image and is not desirable. This can be circumvented by applying a hierarchical dilated convolution or a hybrid dilated convolution [21], wherein after the downsampling operation a different and usually increasing dilation rate is applied. Figure 6.4(d-f) shows the receptive field due to hierarchical dilated convolution avoiding holes in them.

6.3.1   Architecture and Learning

The residual network (ResNet) [22] architecture can help construction of deeper neural networks by circumventing the problem of vanishing gradient. The problem of a vanishing gradient arises since deep layered networks having several multiplication operations would result in close to zero values in backpropagation and thus would be a hindrance toward training deeper networks. ResNet introduces the concept of residual block [23], which introduces skip connections

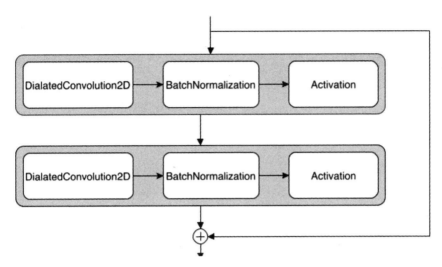

Figure 6.5 Structure of the residual block.

from layer l to the next layer through identity mapping, and thus improves the transfer of gradients during backpropagation.

Figure 6.5 shows the architecture of one residual block.

The block is defined as

$$y = F(x, \{W_i\}) + x; \ i = 1, 2 \tag{6.5}$$

The vector x defines the input and y the output of the ResBlock function (6.5). The residual mapping is defined through the function $F(x, \{W_i\})$ with W_i as the weights of the dilated convolution layers, which are learned during the training. In this case the function F is defined as:

$$F = \sigma \text{Norm}\{W_2 \sigma (\text{Norm}\{W_1 x\})\} \tag{6.6}$$

σ denotes the ReLU activation function and Norm denotes the batch normalization, which is applied after each convolution layer. The dimensions of x and F must be always equal as (6.5) shows. Zero-padding is used to fulfill this condition in the case of different input/output channels or MaxPooling.

Its function (6.5) is originally defined as $y = \sigma (F(x, \{W_i\}) + x); \ i = 1, 2$ with the residual mapping function $F = \text{Norm}\{W_2 \sigma (\text{Norm}\{W_1 x\})\}$.

An exemplary DCNN consists of $3 \times 3 = 9$ ResBlocks separated through two MaxPooling layers. Each ResBlock contains two convolution layers. The net in total consists of $1 + 3 \times 3 \times 2 = 19$ convolution layers.

MaxPooling2D_1 and MaxPooling2D_2 layers with pool shape of (1,2) and stride of (1,2) is applied after ResBlock_1c and after ResBlock_2c, respectively. Each ResBlock contains two dilated convolution layers with the rate two and

Table 6.1
DCNN Architecture [20]

Layer Number	Layer Name	Layer Shape
1	Input_1	(1,8,121)
2	DilatedConvolution2D	(16,8,121)
3	Activation_1	(16,8,121)
4–5	ResBlock_1a	(16,8,121)
6–7	ResBlock_1b	(16,8,121)
8–9	ResBlock_1c	(16,8,121)
10	MaxPooling_1	(16,8,61)
11–12	ResBlock_2a	(32,8,61)
13–14	ResBlock_2b	(32,8,61)
15–16	ResBlock_2c	(32,8,61)
17	MaxPooling_2	(32,8,31)
18–19	ResBlock_3a	(64,8,31)
20–21	ResBlock_3b	(64,8,31)
22–23	ResBlock_3c	(64,8,31)
24	Dense_1	(10)

Table 6.2
Filter Specifications of the Dilated Convolution Layer and
Residual Blocks [20]

Layer Name	Number of Filters	Filter Shape
DilatedConvolution2D	16	(3,3)
ResBlock_1	16	(3,3)
ResBlock_2	32	(3,3)
ResBlock_3	64	(3,3)

zero padding. The specification of the number of filters and the filter shape are shown in Table 6.2. Dilated convolution with a dilated rate of 2 is used for the network. BatchNormalization [24] with a moving average is applied after each dilated convolution layer, which helps the model to converge faster, followed by a ReLU activation function. A fully connected layer with a softmax layer is applied at the end of the ResBlock_3c for classification probabilities. Table 6.2 presents the exemplary neural network architecture with 20 layers used for training the model. The architecture is inspired from ResNet20 except for dilated convolution, some ordering of layers, and operations.

6.3.2 Design Considerations

A binary cross-entropy loss function was used to train the neural network. adam optimizer [25] with a learning rate of 0.0003 was used. Xavier initialization was used as initial weights of the network. Batch size of 128 with 50 epochs were used for training the model.

Image augmentation techniques specifically modeling the sensor artifacts were used to augment the training data set [20]. The following operations were performed on the training data:

- *Random eraser:* Erases part of the range-angle image and sets it to zero. This operation resembles to dropout applied directly to the input image instead of intermediate layers, where it is typically used. However, random eraser mimics the effect of target misdetections in radar and thus is effective in training a model that generalizes well.

- *Random flipper:* Flips the images on the vertical axis. This operation mimics the effect of sensor orientation on the material being flipped.

- *Random brightness:* Adjusts the complete brightness of the image, thus taking into account the fact that the same material is sensed from a different height and thus differing path loss and intensity. This operation also captures the variations due to noise characteristics.

6.3.3 Results and Discussion

The complete training data set has $10 \times 625 = 6,250$ images from all 10 materials. A data augmentation factor of 3 was applied to the training data. The test set consists of 1875 images.

Table 6.3 summarizes the different architectures. It seems that 19 convolution layers have the best accuracy in comparison to a higher amount of layers. The table illustrates the advantages of using dilated convolution and data augmentation along with ResNet for the classification problem.

Table 6.4 presents the confusion matrix of the test set using the ResNet20 model with dilated convolution and trained using image augmentation. The overall classification accuracy is 97.81%. It can be observed that some images of the chocolate cream (b) are predicted as the aluminum surface of a laptop (d). As

Table 6.3
Comparison of Different Approaches and Setups with our DCNN

Approach	Description	Accuracy
DCNN	19 convolution layers	92.91%
DCNN	19 convolution layers with Image Augmentation	93.60%
DCNN	19 dilated convolution layers without Image Augmentation	94.24%
ResNet20	19 convolution layers	96.31%
ResNet20	19 dilated convolution layers with Augmentation	97.81%
ResNet32	31 convolution layers	88.37%

Table 6.4
Confusion Matrix of the Test Set Containing 1,875 Images from All Classes [20]

		Predicted Class								
		a	b	c	d	e	f	g	h	i	j
Actual Class	a	187	0	0	0	0	0	0	0	0	0
	b	1	172	0	11	0	0	0	0	0	0
	c	0	0	189	0	0	0	0	1	0	0
	d	0	0	0	183	0	0	0	0	0	0
	e	0	0	1	0	175	1	0	0	6	6
	f	0	0	0	0	0	199	2	0	0	0
	g	0	0	0	0	0	0	182	0	0	1
	h	0	0	0	0	0	0	4	190	0	7
	i	0	0	0	0	0	0	0	0	159	0
	j	0	0	0	0	0	0	0	1	0	198

The overall accuracy is 97.81%

Table 6.5
Predictions and Accuracy Results for Different Gaussian
Noise Levels [20]

$\sigma^2 =$	0	0.01	0.05	0.1	0.2
Wool	125	125	125	124	96
Other Material	0	0	0	1	29
Accuracy	100%	100%	100%	99.2%	76.8%

well, wood (e) and tile (h) are in some cases classified as water (j). The network requires 2.1E7 FLOPs with 8.0M parameters. The size of the model is 2.53 MB, without quantization, weight pruning, and model fusion.

Further, to test the robustness of the model against sensor artifacts such as variations due to noise, raw ADC for 125 test wool images is taken and then perturbed with additive white Gaussian noise with varying noise levels represented by the variance of the Gaussian noise. Table 6.5 shows the prediction results and accuracy for different variance levels; as can be observed the accuracy remains stable at around 100% until a noise level with a variance of 0.1. This demonstrates the range angle image generated using Capon algorithm is sufficiently robust towards noise artifacts and perturbations, which are necessary for product-ready solutions.

Figure 6.6 presents the range angle feature images of wool for different variance levels of Gaussian noise. In the range angle image the peaks until $\sigma^2 = 0.1$ look similar and only starts becoming affected for a variance level of $\sigma^2 = 0.2$, where the model starts predicting incorrect classifications.

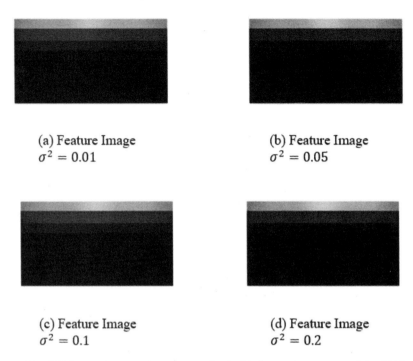

(a) Feature Image
$\sigma^2 = 0.01$

(b) Feature Image
$\sigma^2 = 0.05$

(c) Feature Image
$\sigma^2 = 0.1$

(d) Feature Image
$\sigma^2 = 0.2$

Figure 6.6 RAI feature images of wool normalized with the maximum value of the RAI image without noise for the variance levels of (a) 0.01, (b) 0.05, (c) 0.1, and (d) 0.2 [20].

6.4 Siamese Network

Figure 6.7 depicts five different types of four materials: carpet, tile, liquid, and laminate. In a practical scenario, the model could be trained with the first four types of each material whereas during inference the fifth type of material can be presented for classification. A product-ready model deployed in an industrial, consumer, or automotive context is expected to scale under such scenarios. Further, the material to be classified can have a different background; for example, a vacuum-cleaning robot trying to identify water or liquid on the floor needs to classify liquid under different background conditions, such as liquid over laminate or marble or tile. Thus, the scalability of the model under different backgrounds needs to be ensured. Also, the model should be capable of rejecting alien material, such as silk, which was not part of the training data set.

Although conventional DCNNs can learn robust features from the training data and produce impressive classification accuracies, they require huge amounts of training data that captures all the variations in the data as described earlier. To overcome the bottleneck, one-shot learning can be used to address the above limitations. One-shot learning uses Siamese CNNs to automatically extract robust and discriminative features from different material input images.

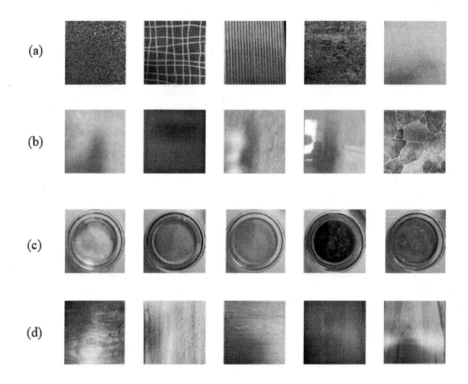

Figure 6.7 Different types (Type I-V) of (a) carpet, (b) tile, (c) liquid, and (d) laminate [19].

A Siamese network employs sister networks with shared weights to learn feature representation to ensure the same/similar materials are grouped together in latent space whereas materials from different classes are widely separated in the latent space. As the Siamese network is trained, the learned discriminative features can be leveraged to generalize the predictive power of the architecture to classify another type of the same materials or reject unknown materials.

6.4.1 Architecture and Learning

A Siamese network architecture consists of two CNNs with shared weights during training. In an example for our classification problem, each CNN has five DilatedConvolution2D, two MaxPooling2D, and two Dense layers with batch normalization layers followed by activation, as described in Table 6.6. The dilated convolution layers [26] of rate two is used. A ReLu activation function was used and the MaxPooling layers uses (1,2) pooling with a stride of (1,2). Similar to earlier architecture, MaxPooling was applied only to the angle dimension and the global field integration is achieved through the dilated convolution layers. The neural networks used in this proposed R-SiameseNet is inspired by LeNet [27].

Table 6.6
R-SiameseNet Architecture [19]

Layer Number	Layer Name	Filter Shape	Layer Shape
1	Input_1		(1,8,121)
2	DilatedConvolution2D_1	(6,6) × 6	(6,8,121)
3	Activation_1		(6,8,121)
4	MaxPooling2D_1		(6,8,61)
5	DilatedConvolution2D_2	(5,5) × 12	(12,8,61)
6	Activation_2		(12,8,61)
7	MaxPooling2D_2		(12,8,31)
8	DilatedConvolution2D_3	(4,4) × 24	(24,8,31)
9	Activation_3		(24,8,31)
10	DilatedConvolution2D_4	(4,4) × 48	(48,8,31)
11	Activation_4		(48,8,31)
12	DilatedConvolution2D_5	(3,3) × 96	(96,8,31)
13	Activation_5		(96,8,31)
14	Flatten_1		(23808)
15	Dense_1		(200)
16	BatchNorm_1		(200)
17	Activation_6		(200)
18	Dense_2		(8)

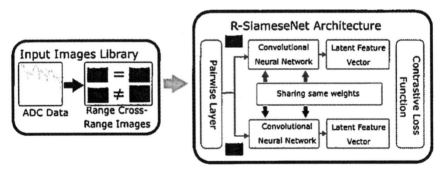

Figure 6.8 R-SiameseNet framework for material classification [19].

6.4.2 Design Considerations

A Siamese network as described in Figure 6.8 learns the process of classification rather than explicit classification, as conventionally done. The twin network [28] is presented with images from the same material class, in which case the objective is to minimize the distance between the two projections in the embedding latent space, and also different material class, in which case the objective is to maximize the distance between the two projections in the embedding latent space. The idea and the objective of the framework is described in Figure 6.8. A latent embedding space with dimension $d = 8$ is used. Note that a larger dimension with and examples can lead to the problem of the curse of dimensionality.

Let X_1, X_2 be a pair of input vectors resulting from two stacked images from the labeled training set. The label y displays if both the input images belong to the same class or to another class. If $y = 1$ the inputs X_1 and X_2 belong to the same class, and if $y = 0$ they are dissimilar class. G_W denotes the output embedding from one of the network in the SiameseNet framework, where W represents the neural network parameters. The parameter C describes the margin value, which is greater than zero, and is used to ensure that dissimilar pairs beyond this margin do not contribute to the loss. The Euclidean distance between the outputs of the neural networks G_W of the both inputs X_1 and X_2 is defined as $D_W(X_1, X_2) = ||G_W(X_1) - G_W(X_2)||_2$. The contrastive loss function, which is used to update the twin network, is then given as

$$L(W, C, X_1, X_2, y) = yD_W(X_1, X_2) + (1 - y)\max(0, C - D_W(X_1, X_2)) \quad (6.7)$$

In the case of two training images labeled to the same class, (6.7) reduces to only the first term where the objective is to minimize the distance between X_1 and X_2. In the case of two different images, the equation reduces to the second term where the objective is to maximize the distance larger than the margin C [29]. The $x_+ = max(0, x)$ operation is used to ensure that this loss doesn't drift toward negative infinity and dominate the overall loss function.

The weights of the convolution layers and dense layers were initialized using Xavier Initialization. Stochastic gradient descent (SGD) with a learning rate of 0.0003 is used. Batch size used is 128 and the network trained over 32 epochs.

6.4.3 Results and Discussion

To evaluate the performance of the R-Siamese network, a data set with five different types of tile, laminate, carpet, and liquid is used with 750 range angle images for each type. Thus, the total data set is $4 \times 5 \times 750 = 15,000$, which is split randomly into 80% to 20% among the training and test sets. The range angle images are generated by enclosing the sensor in a casing by placing it over the materials with a gap of 1–4 cm and in different orientation. The R-Siamese network is trained by randomly sampling images from the same material pairs (i.e., genuine pair) and different material pairs (i.e., imposter pair).

Once the twin network is trained using contrastive loss function, during inference only one of the CNNs is used. Unlike in conventional Deep Net, the CNN doesn't learn to predict the classification probabilities of the input image but learns to project it into an embedding latent space where the similar materials are grouped closer than the dissimilar materials. From the well-learned latent space, either a SVM or k-nearest neighbor (kNN) or dense layer can be used for the final classification. In one case, kNN is used with $k = 10$ nearest neighbors.

Table 6.7
Confusion Matrix of the Test Set Containing 3,000
Images from All Classes and All Types [19]

		Predicted Class			
		Tile	**Carpet**	**Liquid**	**Laminate**
Actual Class	Tile	718	5	1	4
	Carpet	6	751	1	1
	Liquid	0	1	760	0
	Laminate	2	1	1	748

The overall accuracy is 99.23%.

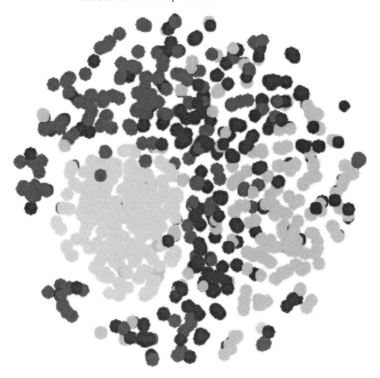

Figure 6.9 t-Distributed stochastic neighbor embedding (t-SNE) representation of the the learned latent feature space of the R-SiameseNet. Different intensity refers to tile, carpet, liquid, and laminate [19].

Table 6.7 presents the confusion matrix of the material classification solution using SiameseNet.

Classification performance with an accuracy of 99.23% is achieved on the test data set. Figure 6.9 shows the t-distributed stochastic neighbor embedding (t-SNE) [30] representation in the 2D space of the learned latent feature space

(a) Normal tile : RAI   (b) Tile with subtle crack: RAI

Figure 6.10   Radar RAI for nondestructive testing applications with (a) normal tile, and (b) tile with subtle crack.

from the training data set, where four clusters are already clearly discernible even in the approximate 2D space.

6.5 Future Work and Directions

Material classification using radar has been demonstrated to work reliably on a prototype system. However, transition to deployable platforms such as mobile vacuum robots or automotive or drones requires vibration cancellation algorithms, which are needed for stabilizing the classification system. The same is true for deployment of material classification system on handheld devices where ego motion cancellation is necessary. Further, for drilling systems for identification of pipes and cables, through-the-wall clutter rejection algorithms are required for sensing and identification of materials beneath the wall. The processing pipeline for material classification can be further extended to nondestructive testing. Figure 6.10 presents the range-angle image of a normal tile and tile with a subtle crack. The concepts of material classification can be easily extended to nondestructive testing applications.

6.6 Problems

1. Define dilated/atrous convolution. How does dilated convolution help in the case of small-sized images?
2. Explain the gridding effect in dilated convolution. How can this effect be circumvented?
3. What is the significance of erasure in an input radar image as a data augmentation technique?
4. What is the rationale behind using Siamese network for material classification?
5. Define contrastive loss function. Derive contrastive loss for a one-class problem such as anomaly detection in nondestructive testing.

References

[1]   Ciuonzo, D., A. De Maio, and D. Orlando, "A Unifying Framework for Adaptive Radar Detection in Homogeneous Plus Structured Interference Part II: Detectors Design," *IEEE Transactions on Signal Processing*, Vol. 64, No. 11, June 2016, pp. 2907–2919.

[2]   Eguíluz, A. G., I. Rano, S. A. Coleman, and T. M. McGinnity, "A Multi-Modal Approach to Continuous Material Identification through Tactile Sensing," IEEE/RSJ International Conference on Intelligent Robots and Systems (IROS), Daejeon, 2016, pp. 4912–4917.

[3]   Baishya, S. S., and B. Buml, "Robust Material Classification with a Tactile Skin Using Deep Learning," IEEE/RSJ International Conference on Intelligent Robots and Systems (IROS), Daejeon, 2016, pp. 8–15.

[4]   Weinmann, M., and R. Klein, "A Short Survey on Optical Material Recognition," in *Proceedings of the Third Workshop on Material Appearance Modeling: Issues and Acquisition (MAM '15)*, Eurographics Association, Aire-la-Ville, Switzerland, 2015, 35–42.

[5]   Varma, M., and A. Zisserman, "Statistical Approaches to Material Classification," ICVGIP, 2002.

[6]   Santra, A., I. Nasr, J. Kim, "Reinventing Radar: The Power of 4D Sensing," *Microwave Journal*, Vol. 61, 2018, pp. 26–38.

[7]   Bystrov, A., E. Hoare, T.-Y. Tran, N. Clarke, M. Gashinova, M. Cherniakov, "Road Surface Classification Using Automotive Ultrasonic Sensor," 30th Eurosensors Conference, EUROSENSORS2016.

[8]   Bystrov, A., M. Abbas, E. Hoare, T.-Y. Tran, N. Clarke, M. Gashinova, M. Cherniakov, "Remote Road Surface Identification Using Radar and Ultrasonic Sensors," in 2014 11th European Radar Conference, IEEE, October 2014, pp. 185–188.

[9]   Kesharaju, M., and R. Nagarajah, "Feature Selection for Neural Network Based Defect Classification of Ceramic Components Using High Frequency Ultrasound," *Ultrasonics*, Vol. 62, September 2015, pp.271–277.

[10]   Tanaka, K., et.al., "Material Classification from Time-of-Flight Distortions," *IEEE Transactions on Pattern Analysis and Machine Intelligence*, Vol. 41, No. 12, December 1, 2019.

[11]   Hasch, J., E. Topak, R. Schnabel, T. Zwick, R. Weigel and C. Waldschmidt, "Millimeter-Wave Technology for Automotive Radar Sensors in the 77 GHz Frequency Band," *IEEE Transactions on Microwave Theory and Techniques*, Vol. 60, No. 3, March 2012, pp. 845–860.

[12]   Gader, P. D., M. Mystkowski, and Y. Zhao, "Landmine Detection with Ground Penetrating Radar Using Hidden Markov Models," *IEEE Transactions on Geoscience and Remote Sensing*, Vol. 39, No. 6, June 2001, pp. 1231–1244.

[13]   Sim, H., S. Lee, B.-Ho Lee, and S.-C. Kim, "Road Structure Classification Through Artificial Neural Network for Automotive Radar Systems," *IET Radar, Sonar & Navigation*, Vol. 13, No. 6, 2019, pp. 1010–1017.

[14]   Brook, A., E. Cristofani, M. Vandewal, C. Matheis, and J. Jonuscheit, "3-D Radar Image Processing Methodology for Non-Destructive Testing of Aeronautics Composite Materials and Structures," 2012 IEEE Radar Conference, May 7–11, 2012.

[15] Yeo, H.-S., G. Flamich, P. Schrempf, D. Harris-Birtill, and A. Quigley, "Radarcat: Radar Categorization For Input & Interaction," In *Proceedings of the 29th Annual Symposium on User Interface Software and Technology*, 2016, ACM, pp. 833–841.

[16] Agresti, G., and S. Milani, "Material Identification Using RF Sensors and Convolutional Neural Networks," ICASSP 2019 - 2019 IEEE International Conference on Acoustics, Speech and Signal Processing (ICASSP), May 12–17, 2019, Brighton, United Kingdom.

[17] Morrison, D., D.-H. Birtill, A. N. Houston, and A. Quigley, "AquaCat: Radar and Machine Learning" for Fluid and Powder Identification," Mobile HCI17 Workshop on Object Recognition for Input and Mobile Interaction, September 4, 2017, Vienna, Austria.

[18] Yeo, H.-S., R. Minami, K. Rodriguez, G. Shaker, and A. Quigley, "Exploring Tangible Interactions with Radar Sensing," *Proc. ACM Interact. Mob. Wearable Ubiquitous Technol.*, Vol. 2, No. 4, December 2018.

[19] Weiß, J., and A. Santra, "One-Shot Learning for Robust Material Classification using Millimeter-Wave Radar System," *IEEE Sensors Letters*, Vol. 2, No. 4, 2018.

[20] Weiß, J. and A. Santra, "Material Classification using 60-GHz Radar and Deep Convolutional Neural Network," 2019 International Radar Conference (RADAR), TOULON, France, 2019, pp. 1-6, doi: 10.1109/RADAR41533.2019.171265.

[21] Wang, P., and G. W. Cottrell, "Understanding Convolution for Semantic Segmentation," arXiv:1702.08502v1, 2017.

[22] He, K., X. Zhang, S. Ren, and J. Sun, "Deep Residual Learning for Image Recognition," IEEE Conference on Computer Vision and Pattern Recognition (CVPR), Las Vegas, NV, 2016, pp. 770–778.

[23] He, K., X. Zhang, S. Ren, and J. Sun, "Identity Mappings in Deep Residual Networks," ArXive-prints, 1603.05027, 2016.

[24] Ioffe, S., and C. Szegedy, "Batch Normalization: Accelerating Deep Network Training by Reducing Internal Covariate Shift," arXiv preprint arXiv:1502.03167, 2015.

[25] Kingma, D. P., and L. J. Ba, "Adam: A Method for Stochastic Optimization," ArXiv e-prints, 1412.6980, 2014.

[26] Fisher, Y., and V. Koltun, "Multi-Scale Context Aggregation by Dilated Convolutions," in CoRR, abs/1511.07122, 2015.

[27] Lecun, Y., L. Bottou, Y. Bengio, and P. Haffner, "Gradient-Based Learning Applied to Document Recognition," in *Proceedings of the IEEE*, Vol. 86, No. 11, November 1998, pp. 2278–2324.

[28] Du, W., M. Fang, and M. Shen, "Siamese Convolutional Neural Networks for Authorship Verification," in Proceedings, 2017. (http://cs231n.stanford.edu/reports/2017/pdfs/801.pdf)

[29] Hadsell, R., S. Chopra, and Y. LeCun, "Dimensionality Reduction by Learning an Invariant Mapping," 2006 IEEE Computer Society Conference on Computer Vision and Pattern Recognition (CVPR'06), New York, 2006, pp. 1735–1742.

[30] Maaten, L.V., and G. Hinton, "Visualizing Data Using t-SNE," *Journal of Machine Learning Research*, November 2008, pp. 2579–2605.

7

Vital Sensing and Classification

Avik Santra, Muhammad Arsalan, Christoph Will

7.1 Introduction

Heart rate and breath rate are two common vital signs used for diagnoses of a patient in heath care. Continuous monitoring of these vital signs can lead to better diagnosis by early detection and prevention of critical states of health [1]. While most vital sign detection techniques are based on wearable devices, touch-free detection methods like radar systems offer a comfortable and ubiquitous solution to the user. Radars can detect and sense human cardiopulmonary motion remotely from a distance. In [2–17], radar is demonstrated as a powerful sensor in short-range localization and vital-sign tracking within consumer electronics, medical care, surveillance, driver assistance, and industrial applications. Monitoring the vital signs of human targets finds applications in patient monitoring in hospitals, sleep apnea detection and presence sensing in smart homes, driver monitoring in autonomous cars, and physiological monitoring in surveillance and earthquake rescue operations [4–10]. The signal shape of a single heartbeat is dependent on the subject, the chosen measurement spot, and the distance to the antenna, which increases the complexity and impedes a high accuracy [11].

The conventional approach to extract these vital signals rely on a radar interferometric approach, which further utilizes the discrete Fourier transform on a bandpass-filtered signal and determines breath and heart rates by detecting the maximum peaks in the frequency spectrum [5,12–19]. However, a FFT-based approach requires a large observation window of 10–30 s to achieve reliable frequency resolution, and thus alternative time-domain based approaches have been proposed in the literature. Time domain-based algorithms also enable the detection of individual heartbeats and are more sensitive to heart rate variability.

Exemplary algorithms use feature-based correlation [20], continuous-wavelet filter, and ensemble empirical mode decomposition in combination with a zero-crossing detector [21], linear demodulation to extract the heartbeat signal, and subsequent autocorrelation with statistical analysis to determine the heart rate [22] or (advanced) template-matching investigating up to five heterogeneous heartbeat shapes [23,24].

However, such approaches have several limitations in nonlaboratory measurement scenarios, including artifacts due to multiple reflections, vital signal corruption due to random body movements (RBM), human motions, and inaccuracies arising due to intermodulation products (IMP) or harmonics. Most of the proposed methods in the literature cancel only large, radial movements [25,26] or focus on breath rate [27]. More sophisticated methods detect vital signs in the corresponding time-domain signals [24,28] or eliminate stationary clutter in hardware [29,30], for instance. Recently, an algorithmic clutter elimination method was published that simultaneously suppresses harmonics [31]. Another approach to cancel out the harmonics analyzes the skewness and standard deviation of the collected impulses of an ultrawideband radar [32]. RBMs are canceled using two synchronized radar systems [33], an additional camera [34] or a self- and mutual-injection locking in the RF part [35]. On the software side, RBMs are compensated by complex signal processing methods [25,26,36] or a deep neural network [27]. Deep neural networks allow for general body movement cancelation, but require a large amount of training data and imply a huge computational effort [27]. Moving average filtering only compensates for short-term movements [37].

The aim of practical radar-based vital signal detection deployed in industrial, consumer, and medical applications are to achieve a high accuracy within a short observation window, while eliminating the perturbations and effects due to human motion and RBM and inaccuracies due to IMPs and harmonics. Thereby, in principle the objective of such systems is to sense heart rate or breath rate accurately with high fidelity without jumps or drifts in the estimates.

In this chapter, we introduce the fundamentals of radar-interferometry-based vital signal detection and sensing using common approaches based on FFT, time domain peak search, and template-matching in Section 7.2. Next in Section 7.3, we present a deep-learning-based approach to estimate the vital signal from bandpass filtered radar IQ data, and present their challenges and limitations. We then in Section 7.4 present a novel vital signal Kalman filter tracking approach with adaptive filtering of the time domain radar signal to focus and track heart rate and demonstrate that the solution can achieve much better results compared to state-of-the-art approaches. We subsequently present the shortcomings of the proposed approach and outline deep-learning-based methods to overcome such challenges. Thus, a hybrid solution involving both signal processing and deep learning algorithms is presented to continuously estimate and track vital signal eliminating effects due to motion, RBM, and IMPs in Section 7.5. The binary classifier acts as a filter with a brain that passes an IQ signal only when it knows the estimation through subsequent approach would be accurate. The functionality of the presented solutions is demonstrated using Infineon's 60-GHz FMCW *BGT60TR13C* radar chipset while comparing the measurement results with the state-of-the-art algorithm in reference to a commercial chest belt. We conclude with future research directions in Section 7.6.

7.2 Vital Signal Fundamentals

FMCW radar transmits a linearly increasing frequency waveform, called chirp, which after reflection by an object is collected by a receiver antenna. Afterward, both the transmitted and received signal are mixed together by a mixer at the receiver and the resultant lowpass filtered signal is called the intermediate frequency signal.

The frequency of the FMCW waveform with bandwidth B and duration T can be expressed as:

$$f_T(t) = f_c + \gamma t, \tag{7.1}$$

where f_c is the ramp start frequency and $\gamma = \frac{B}{T}$ is the chirp rate. The reflected signal from the target is mixed with a replica of the transmitted signal, resulting in a beat signal. The phase of the beat signal after the mixer due to the target at base location R from the sensor is:

$$\Phi(t) = 2\pi \left(\gamma t \tau + f_c \tau - \frac{\gamma}{2} \tau^2 \right) \tag{7.2}$$

where $\tau = \frac{2R}{c}$ is the round-trip propagation delay between the transmitted and received signal after reflection from a static target.

Consider Figure 7.1, where the time evaluation of a vibrating source, such as a human heart, is shown. The source is deviating to the left and right with a small displacement due to the vibrating source. Now when a radar is placed in front of this vibrating source and multiple consecutive chirps are sent, each of reflected chirp will have a peak in the same location but with different phase. Thus, the measured phase of the peak across time exhibits a sinusoidal periodic pattern, which can be measured to estimate the vibration frequency.

The movement of the thorax of the human body can be modeled as a vibrating source as shown in Figure 7.2. The movement of thorax due to breathing and heartbeat has unique patterns. For more details about the cardiovascular physiology readers can refer to [38]. The displacement due to this human chest movement results in a change of phase over time and is referred to vital-Doppler or micro-Doppler in the literature. In this chapter, we use vital-Doppler to refer to this motion due to breathing and heart rate.

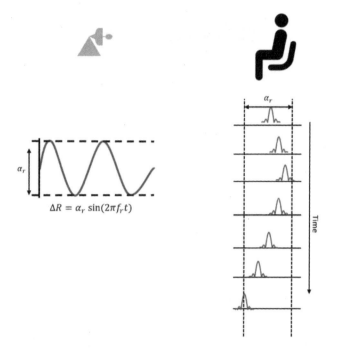

Figure 7.1 Human heart as a vibrating source.

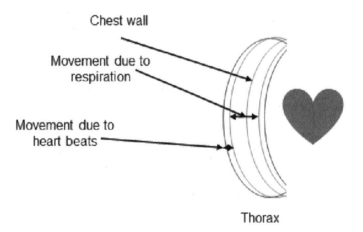

Figure 7.2 Displacement of human thorax due to breathing and heartbeats.

Due to vital-Doppler, the range profiles of a human target at different measurements have significant variation, whereas a the range profiles for stationary objects are very stable. Taking the standard deviation of these measurements will result in a peak at the location of the human subject and thus a human target can be separated or differentiated from other stationary targets [39]. As well, the signals obtained for these vital signals can be further processed through signal processing approaches to estimate the heart rate and respiration rate separately.

Therefore, heart rate and respiration rate can be estimated on filtering the extracted IQ phase signal through appropriate bandpass filters [19]. The detected range response of the human target at a distance R from the center of the radar arising due to the vibrating sources can be expressed as:

$$R(t) = R + \alpha_r \sin(2\pi f_r t) + \alpha_h \sin(2\pi f_h t) \qquad (7.3)$$

where the reflections from the human thorax can be modeled as a superposition of two primary vibrating sources: one vibration due to respiration and another due to heartbeat. The human resting respiratory rate is around 12–24 beats per minute (bpm); that is, $f_r = 0.2 - 0.4$ Hz with a maximum displacement of the chest wall $\alpha_r = 7.5$ mm and the heart rate can be from 48–200 bpm; that is, $f_h = 0.8 - 3.33$ Hz with a maximum displacement of $\alpha_h^{(2)} = 0.25$ mm. Therefore, the heart rate and respiration signals can be estimated on filtering the extracted IQ phase signal through appropriate bandpass filters.

Substituting $R(t)$ into (7.2), the received phase signal can be expressed as:

$$\phi(t) = 4\pi \left(\overbrace{\frac{\gamma R(t)}{c}}^{\text{1st term}} t + \overbrace{\frac{f_c R(t)}{c}}^{\text{2nd term}} - \overbrace{\frac{\gamma R^2(t)}{c^2}}^{\text{3rd term}} \right) \qquad (7.4)$$

The propagation delay of the round trip in this case of a vibrating source is given as:

$$\tau = \frac{2R(t)}{c}. \qquad (7.5)$$

The second-order undesired terms in (7.4), such as $f_r + f_h$, referred to as intermodulation products and $2f_r$, referred to as harmonics fall within the vital signal spectra. If not considered, these undesired terms lead to inaccurate estimates of the vital signals [40] and are a source of estimation error. The other terms in (7.4) lead to the expression:

$$\phi(t) = 4\pi \left(\left(\overbrace{\frac{\gamma R}{c}}^{\text{1st term}} + \overbrace{\frac{\gamma \alpha_r \sin(2\pi f_r t)}{c}}^{\text{2nd term}} + \overbrace{\frac{\gamma \alpha_h \sin(2\pi f_h t)}{c}}^{\text{3rd term}} \right) t \right.$$

$$\left. + \overbrace{\frac{f_c R}{c}}^{\text{4th term}} + \overbrace{\frac{\alpha_r f_c \sin(2\pi f_r t)}{c}}^{\text{5th term}} + \overbrace{\frac{\alpha_h f_c \sin(2\pi f_h t)}{c}}^{\text{6th term}} \right). \qquad (7.6)$$

The first term is estimated by fast-time FFT and is proportional to the range of the human thorax R. The second and third term in (7.6) can be ignored by considering $R \gg \alpha_r, \alpha_h$. The fifth and sixth terms are the change in phase over slow time due to respiratory and heartbeat motion generated from the human thorax wall. The fourth term is a fixed-phase offset due to the base distance R. Monitoring the phase followed by spectral filtering enables the estimation of the nonoverlapping heartbeat and breathing frequencies.

The presented theory is true for ideal circumstances; however, due to the sensitivity of the phase to very minute motion, the estimates of heartbeats and breathing frequencies get corrupted by random body movements and human motion, and therefore estimating heart or breathing rate in practical unconstrained circumstances is a challenge and requires algorithmic innovations. Further, as outlined earlier, the second-order terms in (7.2) result in intermodulation products, $f_h + f_r$, and harmonics, $2f_r$, which on several occasions dominate the time segments, leading to incorrect heart-rate estimates if not handled appropriately.

7.2.1 Preprocessing Steps

FMCW radar systems, which are used to detect targets with high speed, generally process radar signals through 2D range-Doppler maps and transmit PN consecutive chirps with a short PRT among them in a frame. The short PRT implies larger unambiguous velocity estimation, which is essential in the case of high-speed targets, and range-Doppler maps help in uncoupling the range and Doppler uncertainty exhibited due to LFM's ambiguity properties.

On the other hand, for pure vital sensing applications, the FMCW radar system requirements are different. While the PRT does not need to be extremely short, typically a larger observation window is preferred (it enables good frequency resolution in the case of FFT-based approaches). But more importantly, the time interval between the last chirp of a frame to the beginning of the first chirp of the consecutive frame has to be identical to the PRT as all chirps have to be transmitted (and received) equidistantly within the observation window, which at times can be longer than the HW's frame interval. Thus, the effective sampling rate of the vital-Doppler data in each range bin is 1/PRT. An example of such processing is depicted in Figure 7.3, wherein the range-cross range imaging is performed within the radar's physical frame boundary, while the first chirp from consecutive frames is extracted to build the observation window for the vital signal processing. Nonetheless, the update rate of the vital signal processing can be enhanced through a sliding window mechanism of the observation window. Furthermore, vital signal processing relies on monitoring the IQ phase data, also referred to as the radar interferometric technique, and doesn't necessarily require generating range-Doppler maps.

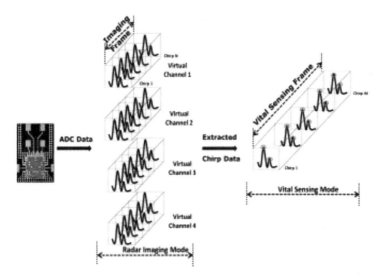

Figure 7.3 Extraction of vital frame boundary from multiple physical frames ensuring equidistant PRT between extracted chirps [19].

The conventional pipeline to process the vital signal through interferometry technique involves target detection through range FFT applied over all chirps in the frame. The range bins of the potential targets can be determined by either of the following approaches:

- *Peak detection in range FFT*: The summation of complex FFTs within the vital observation window is taken (coherent integration), followed by peak search on the amplitude of the summed FFT spectrum;

- *Peak-to-average ratio (PAPR)*: The ratio of peak to average across the slow time of each range bin from the FFT spectrum is calculated. For a stationary target, the peak FFT value is close to the average/mean of FFT spectrum along the slow time, whereas in the case of a vibrating source the mean/average value would be close to zero, resulting in a high PAPR.

In the next step, following target detection along range, a vital Doppler detection is performed by either

- Estimating the standard deviation of the IQ data across slow time and checking if they lie within a prescribed value;

- Vital Doppler FFT across the vital observation window, followed by checking if the signal peak within the vital frequencies (0.2–3.3 Hz) crosses a prescribed threshold.

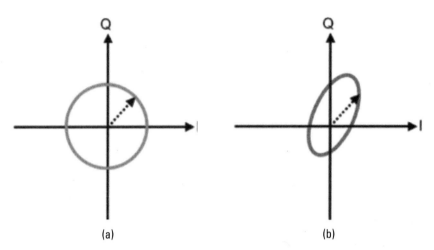

Figure 7.4 (a) Exemplary ideal IQ plot, and (b) exemplary real IQ plot due to nonidealities in the hardware.

The vital-Doppler detection is an important step before passing the signal through vital bandpass filters to eliminate false static target bins since white noise will lead to valid incorrect signal if passed through vital bandpass filters.

Following the vital detection step, the IQ data of the range bins that pass the above criteria is fitted with an ellipse reconstruction algorithm to compensate for offset, phase, and amplitude imbalances due to hardware imperfections. Since the signal is of a complex nature, if plotted on a complex plane, ideally it should look like a unit circle as shown in Figure 7.4(a), where I is the in-phase part of the signal and Q is the quadrature part. But due to hardware imperfections, the resulting signal on the IQ plot looks like an ellipsoid, as shown in Figure 7.4(b) and often with a bias, where the center is not around 0. The goal of the ellipse reconstruction is to remove these amplitude and phase offsets by mapping the ellipse onto a unit circle [41].

The compensated data is then fed to the angle calculation block where the angle or phase is calculated by arctangent demodulation. The I and Q parts of the signal are given by:

$$I = A\cos(\phi),$$
$$Q = A\sin(\phi). \tag{7.7}$$

From the above equation, the amplitude A and phase shift ϕ is obtained as:

$$A^2 = I^2 + Q^2,$$
$$\phi = \arctan(Q/I). \tag{7.8}$$

The resulting phase is then passed through the phase unwrapping block which ensures that all relevant multiples of 2π have been included in the phase. Phase unwrapping is a process of reconstruction of the original true phase of a wave from its multiples of 2π values. This phase is unwrapped by adding or subtracting 2π for phase jumps larger than $-\pi$ or $+\pi$, respectively. With λ being the wavelength of the carrier frequency and $\lambda/2$ representing the unambiguousness (phase) range, the displacement of the target can subsequently be calculated by the following formula:

$$\Delta d = \frac{\lambda}{4\pi} \cdot \text{unwrap}\,(\phi) \tag{7.9}$$

The resultant displacement signal contains the information of breathing rate and the heart rate as presented in (7.6). Therefore, the displacement signal is fed through bandpass filters, with start and stop frequencies as 0.2 Hz and 0.4 Hz, respectively, for breathing rate estimation and 0.8 Hz to 3.3 Hz as start and stop frequencies for heart rate estimation.

Finally, the frequency of the heart rate or breathing rate can be calculated by several methods. Some of the possible approaches are outlined here:

- *Spectral estimation techniques through FFT of the filtered and compensated vital signal*: The peak in the FFT spectrum within the heart rate frequency band and breathing rate frequency band depicts the heart rate and breathing rate estimates respectively.

- *Estimate the breathing/heart rate by counting the peaks detected in the filtered time-domain vital data in the observation window*: Utilizing constraints like minimum peak height and/or minimum peak prominence prevents the detection of too many peaks per heartbeat, if higher frequent pulse wave components are in the heartbeat signal. The more intelligent the peak search, the better the performance.

- *Template-matching the time-domain vital data with templates from a saved library by cross-correlation*: While template-matching can be performed with only one template, the performance increases when more templates are used. In heartbeat detection, heterogeneous template types represent the variety of the detectable pulse wave forms depending on the measurement spot and the antenna's field of view. Homogeneous templates of each type then represent the small changes of the heartbeat shapes over time or between different persons. The choice of the template type to be used for the current vital signal under investigation can be made by feature detection for instance. Breath rate and heart rate can subsequently be estimated by counting the peaks in the cumulative cross-correlation domain within the observation window.

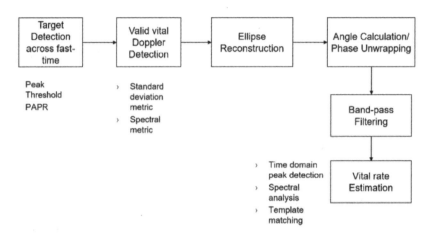

Figure 7.5 Standard vital signal processing pipeline.

Figure 7.5 presents a standard state-of-the-art processing pipeline for vital signal extraction and estimation using several of the standard approaches in FMCW radar.

7.3 Heart Rate Estimation through a Deep-Learning Approach

Recent advances in deep learning have enabled major breakthroughs in problems that were earlier not solvable through conventional or model-based approaches. In this section, we explore different DL methods to estimate the heart rate or breathing rate through classification of the IQ signal, while rejecting inaccuracies arising from RBM, human motion, and IMP. The deep learning models are required to learn the nonlinear relationship between the expected features for accurate heart rate estimation and the input IQ signal.

Once the target range bin is identified and the IQ phase signal is compensated and filtered using approaches described in the earlier section, after which the signal is fed into the deep neural network for the classification or regression as shown in Figure 7.6. The signals are labeled using a *Polar H10* chest belt as ground truth.

The data is collected for many subjects using the following procedure. The data volume obtain from the radar is in form of $[N_f, N_r, N_{ts}]$ where N_r is the total number of chirps, N_f is the total number of frames, and N_{ts} is the total number of samples per chirp. The parameters and other system parameters used for collecting the measurement data are presented in Table 7.1.

The subjects were placed in front of the radar at a distance of approximately 40–60 cm as depicted in Figure 7.7 without any supervised restriction on random body movements. The commercial chest belt *Polar H10* was used as a heart rate reference sensor. Fourteen human healthy subjects participated in the

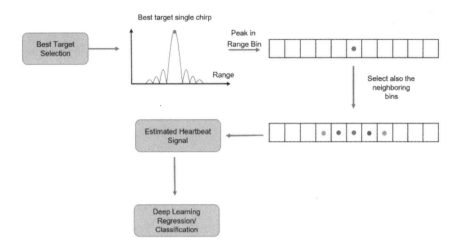

Figure 7.6 Deep learning approach.

Table 7.1
System Operating Parameters

Parameters	Symbol	Value
Operational frequency	f	60
Pulse repetition time	PRT	5 ms
Ramp start frequency	f_{min}	59 GHz
Ramp stop frequency	f_{max}	61 GHz
Bandwidth	B	2 GHz
Sampling frequency	fs	0.747664 MHz
Number of samples per chirp	N_{ts}	256
Frame time	Fr_t	1.28 s

experimental results. For half of the subjects, a measurement row with common breathing and for another half, a measurement row with breath-holding was recorded. All measurement rows have a length of 16 frames, which corresponds to approximately 20 s.

Then, these data is divided in to different classes in such a way that for the signals whose reference values are in a range of 46–55 bpm, which corresponds to 0.75 Hz–0.83 Hz are labeled as class 1, signals having reference values in a range of 51–55 bpm are labeled as class 2 up to class 12, which contains the signals having reference values in a range of 96–100 bpm corresponding to 1.5 Hz–1.6 Hz. The measurement campaign involved collecting data from humans under the following circumstances to include data variations:

1. Without any movement and with normal breathing;
2. Without any movement and without any breathing;

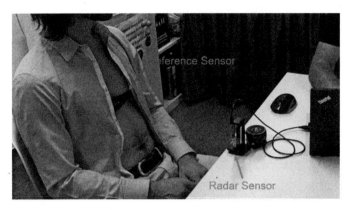

Figure 7.7 Experimental setup [42].

3. With random body movements and with normal breathing;
4. With random body movements and without breathing;
5. Run/jog and then without any movement and with normal breathing;
6. Run/jog and then without any movement and without breathing;
7. Run/jog and then with random body movements and with normal breathing;
8. Run/jog and then with random body movements and without breathing.

The intention of measurement post-running/jogging is to capture heart rates corresponding to higher labels.

Figures 7.8 and 7.9 present examples of variations in the radar heart rate time-domain signal for the class of 61–65 and 66–70, respectively.

7.3.1 GAN-Based Data Augmentation

Once the data was collected, it was observed that some frequency classes had large data examples, while other classes had very few example data (especially at higher-frequency bands), which could introduce bias to the training of the deep learning model. Thus, to circumvent such issues, GAN-based data augmentation is proposed and implemented. The GAN-based data augmentation approach feeds the generator with random numbers while the discriminator is fed with real data, which corresponds to the radar sensor's data and fake data, which is the output of the generator. The GAN-based training data augmentation scheme is shown in Figure 7.10. The two models are trained together in a zero-sum game, adversarial, until the discriminator model is fooled about half the time, meaning the generator model is generating plausible examples.

The generator network takes an input of 256 × 1 signals followed by a dense layer with 256 outputs with a leaky ReLu activation. The dense layer is followed by three convolution layers with a number of filters of 256 each and filter sizes of 3, 5, and 7, respectively. All the convolution layers take Leaky Relu

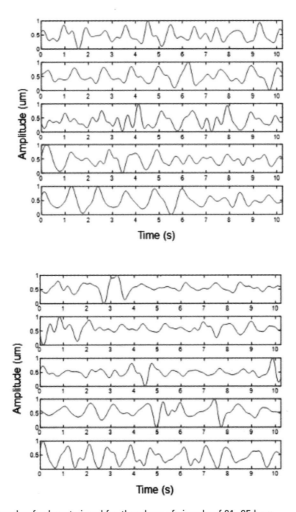

Figure 7.8 Examples for heart signal for the class of signals of 61–65 bpm.

as an activation function except the last convolution layer, which takes tanh as an activation function.

The discriminator network takes an input of 256 × 1 signals followed by two convolution layers with a number of filters of 64 and filter size of 3 and stride 2. Each convolution layer takes leaky ReLu activation functions and is followed by a dropout layer during training. The last layer of the discriminator network is a dense layer with binary class output (i.e., real or fake).

An Adam optimizer is used for both the discriminator network and generator network, which computes the adaptive learning rate for each network weight over the learning process from estimates of the first and second moments of the gradients. In configuration parameters, the learning rate (alpha) is set to 0.0002

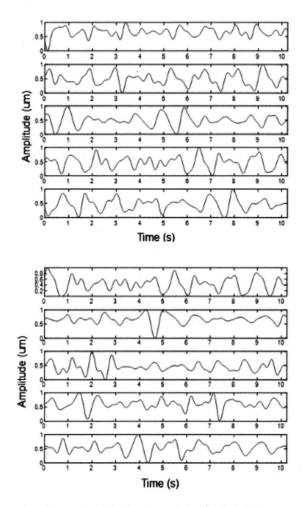

Figure 7.9   Examples of heart signal for the class of signals of 66–70 bpm.

and the exponential decay rate for the first (beta1) and second (beta2) moment estimates are set to 0.5 and 0.999. The epsilon that counters divide by zero problems is set to 1e-8.

The GAN network is able to learn the distribution and to produce plausible radar signal examples. For instance, one of the synthetic data that is generated using the GAN network is shown in Figure 7.11.

7.3.2   Results and Discussions

Classification accuracy is used as a measure of performance of the networks. The results are shown in Table 7.3. It can be seen that the DL methods with different data transformations could achieve the best accuracy up to 55%–60% on

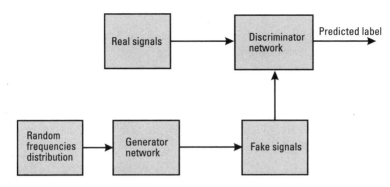

Figure 7.10   GAN model for data augmentation.

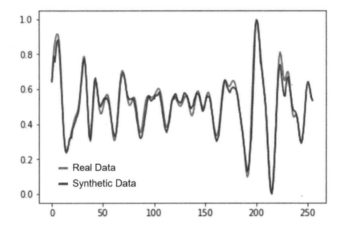

Figure 7.11   Example of a synthetic signal for augmentation.

test/unseen data and hence failed to work for this problem. Similarly, conventional ML algorithms like random forest and SVM also fail to classify and even perform worse compared to DL methods. In the case of the deep learning approach, several adjacent range bins to the one where a target was detected is also fed as input channels and only the best target range bin in other cases. Further, only using the IQ data directly as input to the Deep Net, various transformations, such as wavelet transform using Morlet wavelets, FFT spectrograms, one-hot encoding of the FFT peaks, and PCA on the input data were tried. However, none of the transformations or preprocessing steps were able to provide accuracies better than 55%–60%.

When the augmented data generated with GAN is added to both training and test data sets, the accuracy reached 88%. However, when testing is performed only on the real test set, without including any augmented data, the accuracy drops down to 40%. This drop in accuracy is due to the common pattern introduced by the GAN augmented data and artifacts not present in the real data.

Table 7.2
Classification Results of Different DL and ML Methods, Including Various
Pre-processing and Transformations

Deep Learning Architectures	Pre-processing	Transformations	Accuracy
LSTM BiLSTM Stacked LSTM CNN-LSTM ConvLSTM CNN	Only Best Target Best Target + Neighboring Bins	Wavelet Maps One Hot Encoding FFT Spectrograms PCA	55–60%
ML Methods Random Forest SVM			45–50%

Table 7.3
Performance Evaluation

Test Data	Accuracy
Test Data + Augmented Data	88%
Only Test Data	40%

Thus, deep learning methods fail to achieve good classification accuracy for the complex problem of classifying a heart motion signals while removing the RBM and IMP effects. This indicates that there are no common patterns in the signals, which, current state-of-the-art deep learning models are capable of learning and classifying. Further, the GAN-based approach overfits the data set, leading to poor classification accuracies.

In the next section, we present a signal processing based approach utilizing the Kalman filter and adaptive filtering to achieve higher accuracy compared to conventional approaches.

7.4 Adaptive Signal Processing with a Tracking Approach

The algorithm proposed in this section pursues the objective of accurate heart rate estimation in the presence of RBM, IMP, and harmonics by adaptively filtering the time domain heartbeat signal depending on the current heart rate of the subject and tracking the vital signal through a Kalman filter. The bandpass filter limits are not fixed, as in a standard algorithm, but continuously adjusted during each measurement row. The determined rate per time instance is stabilized and smoothed by applying a Kalman filter, which additionally updates the bandpass filter limits. Further, the IQ signal is evaluated for deviation from circularity. If the deviation is less, the signal estimates are allowed to update the Kalman filter; otherwise, else the filter relies only on the Kalman filter prediction.

7.4.1 Algorithm

A flow chart of the main part of the proposed algorithm is shown in Figure 7.14. In the first step of the signal processing loop, the displacement signal as the output of the preprocessing steps is bandpass filtered with a fourth-order Butterworth filter to extract the heartbeat signal. A peak search considering a minimum peak prominence is applied to detect the individual heartbeats, which are subsequently used to calculate a new temporary heart rate value. Initially, the passband range is set to $0.7 \dots 3.0$ Hz, which corresponds to a heart rate of $42 \dots 180$ beats per minute (bpm). The IQ signal for the vital observation window is checked for deviation from circularity. If the deviation is above a predefined threshold like in Figure 7.12(a), it is assumed that the signal has been corrupted due to RBM or human motion and thus the time segment is ignored. In contrast, if the deviation from circularity is less than the threshold as in Figure 7.12(b), high-quality IQ data (HQD) is assumed and the time segment is used for heart rate estimation. The threshold controls the sensitivity of what defines high-quality IQ data, and typically needs to be chosen empirically.

In the case of high quality target data, the filter settings are updated after state prediction and a gating procedure of the Kalman filter. The modified algorithm step of filtering out poor quality data is depicted in Figure 7.13 to include the target evaluation block. If new high-quality target data has been acquired in a frame, the filter limits are adjusted depending on the previously and currently detected heart rates; otherwise, the passband range stays the same. In principle, the aim is to narrow down the passband range to suppress the higher frequency components of the pulse wave in the heartbeat signal [11] and the adverse effects

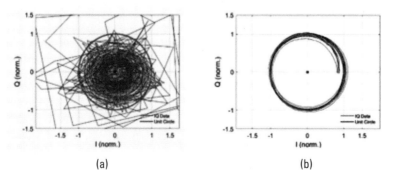

(a)   (b)

Figure 7.12 Exemplary IQ plots of (a) low, and (b) HQD [42].

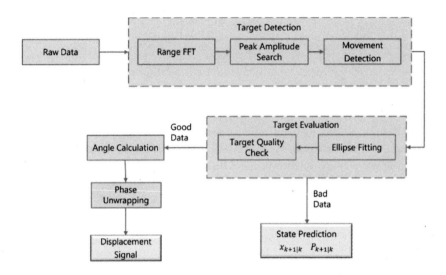

Figure 7.13 Modified flowchart of the preprocessing steps [42].

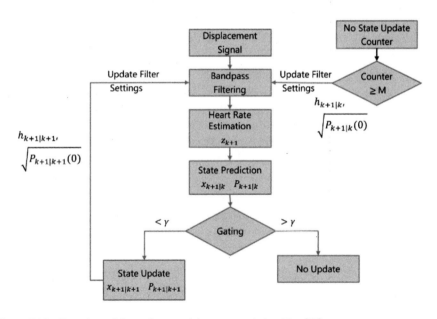

Figure 7.14 Flowchart of the main part of the proposed algorithm [42].

of IMP and respiration harmonics. Furthermore, when the data isn't good quality only the Kalman filter prediction is done, which is based on the underlying heart rate acceleration model, which is explained in details next.

Kalman filtering is a recursive Bayesian process, which is applied when the measured values contain unpredictable or random errors, uncertainties, or variations. The state vector is expressed as:

$$x = \begin{bmatrix} hr & \dot{hr} & \ddot{hr} \end{bmatrix}^T \tag{7.10}$$

where $hr, \dot{hr}, \ddot{hr}$ represents the heart rate in Hz, its first-order derivative, and second-order derivative, respectively. The state vector in the Kalman filter at frame time k is represented as a Gaussian process with mean $x(k|k)$ and covariance or uncertainty $P(k|k)$.

In the algorithm, the Kalman filter is designed as constant acceleration model, and thus the state transition model predicting the state in frame $k + 1$ from state vector in frame k is given as:

$$x(k + 1|k) = Ax(k|k) = \begin{bmatrix} 1 & \Delta t & 0.5\Delta t^2 \\ 0 & 1 & \Delta t \\ 0 & 0 & 1 \end{bmatrix} x(k|k) \tag{7.11}$$

in which Δt denotes the time elapsed between two frames. The observation model is provided as:

$$y(k + 1) = Hx(k + 1|k) = \begin{bmatrix} 1 & 0 & 0 \end{bmatrix} x(k + 1|k) \tag{7.12}$$

The measured heart rate estimation value from the peak count algorithm is denoted as $z(k+1)$. The uncertainty of the state prediction originates from process noise and is modeled as $Q = GG^T \rho_a^2$, where G is $[0.5\Delta t^2 \quad \Delta t \quad 1]$ and ρ_a^2 represents the acceleration process noise. The variance of the measurement noise is denoted as $R = \delta h^2$, which represents the expected square of the estimation error.

Once the first vital data is recorded, a track is initialized with default Q, R noise matrices and initial state and covariance $x(0|0)$, $P(0|0)$ along with default filter settings, the proposed algorithm [42] operates at each frame k as follows:

1. *Bandpass filtering*: The preprocessed displacement signal is a bandpass filter whose cutoff frequencies are defined by the state update from an earlier frame.
2. *Peak count estimation*: The heart rate estimation z_{k+1} is obtained in time domain through peak counts.
3. *State prediction*:

$$x(k + 1|k) = Ax(k|k)$$
$$P(k + 1|k) = AP(k|k)A^T + Q \tag{7.13}$$

4. *Ellipsoidal gating*:

$$y(k + 1) = Hx(k + 1|k)$$

$$(z(k + 1) - y(k + 1))^T \tilde{P}(k + 1|k)^{-1}(z(k + 1) - y(k + 1)) > \gamma_{th}$$

$$(7.14)$$

The ellipsoidal gating function checks if the heart rate estimate $z(k + 1)$ is within the gating window γ_{th} if it is outside the ellipsoidal window, then no Kalman filter update is done as the estimated data is an outlier. If it is within the window, the Kalman filter is updated.

5. *State update*:

$$G(k + 1) = P(k + 1|k)H^T(HP(k + 1|k)H^T + R)^{-1}$$

$$x(k + 1|k + 1) = x(k + 1|k) + G(k + 1)(z(k + 1) - y(k + 1))$$

$$P(k + 1|k + 1) = (I - G(k + 1)H)P(k + 1|k) \qquad (7.15)$$

G(k+1) is the adaptive Kalman filter gain at time step $k + 1$, which determines the weight to apply on the present estimate $z(k + 1)$.

6. *Update filter settings*: Based on the updated state, the bandpass filter settings are updated as start frequency $hr(k + 1|k + 1) - \sqrt{P(k + 1|k + 1)(0)}$ and stop frequency as $hr(k + 1|k + 1) + \sqrt{P(k + 1|k + 1)(0)}$, where (0) represents the first row/column element of the matrix.

If the Kalman filter state update is not done for M consecutive frames, either because of poor-quality data or estimation outside the gating window, the filter settings are updated using the state predictions $hr(k + 1|k) - \sqrt{P(k + 1|k)(0)}, hr(k + 1|k) + \sqrt{P(k + 1|k)(0)}$. This helps in translating the center frequency and also increasing the bandpass filter cutoff settings to account for the fact that the correct heart rate might have drifted during the period there were no updates. The track is killed and filter settings reset once there is no target detections for N consecutive frames. For the demonstrator, M is set to 5, whereas N is set to 10. Figure 7.14 presents the algorithm after the displacement signal is extracted once it is determined that the time segment possesses high-quality data.

7.4.2 Results and Discussion

The performance of the proposed algorithm is evaluated in Figure 7.15, in which each line represents one measurement row. The triangle lines depict the reference values, while the circle lines represent the output of the pure peak detection with

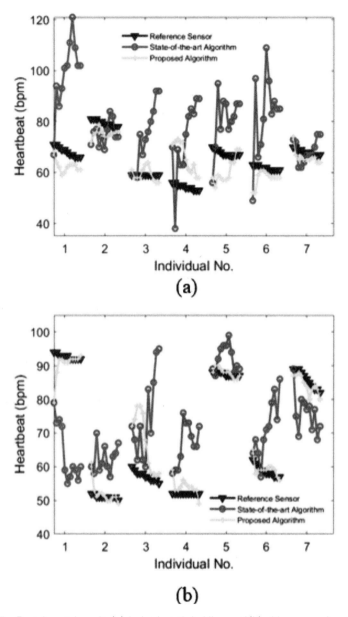

Figure 7.15 Experimental results (a) during breath-holding, and (b) with common breathing [42].

fixed (initial) passband settings based on [19], and the polygon lines depict the results of the presented algorithm.

The presented algorithm achieves better results than the spectral estimation-based algorithm [19] in all measurement rows, as illustrated in Figure 7.15(a).

The rationale behind choosing the spectral estimation-based algorithm is that the base signal processing blocks are aligned with the proposed algorithm. The heart rate values for fixed passband settings show a high deviation from the reference sensor and sometimes have large jumps between two consecutive values. Narrowing down the bandwidth and tracking the heart rate value prevents such jumps and smooths the estimated values. While the first values show a considerable deviation, the Kalman filtering approximates the value toward the reference value step by step. The resulting RMSEs for the presented algorithm are 5.3 bpm and 7.0 bpm compared to 17.6 bpm and 21.3 bpm obtained with a spectral estimation-based algorithm for common breathing and breath-holding cases, respectively, indicating the proposed algorithm outperforms the spectral estimation-based method by a factor of three.

The effect of narrowing down the passband bandwidth is illustrated in Figure 7.16. During the presented time window, the human subject had a heart rate of 59 bpm. In the upper subplot, the filter limits are set to the initial settings, which leads to too many detected peaks and therefore a too-high heart rate of 117 bpm. Narrowing down the heart rate to the minimum passband range of 0.3 Hz, one peak per heartbeat is detected and therefore a quite accurate heart rate of 58 bpm is estimated. If the first estimates with larger bandwidth are wrong, the Kalman filter approximates to a different frequency component, which explains the moderate results of some measurement rows. These frequency components are either harmonics, intermodulations of the breathing signal, or higher frequent pulse wave components with similar amplitude.

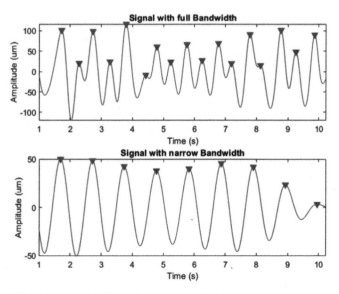

Figure 7.16 Effect of narrowing down the bandwidth [42].

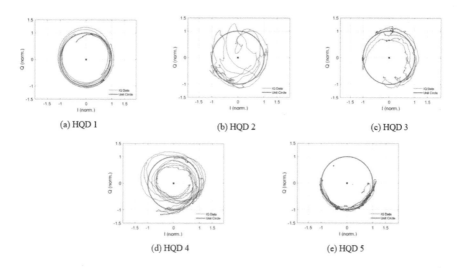

Figure 7.17 Exemplary IQ plots of HQD, the large variations illustrating in the IQ data in terms of size, orientation, and patterns.

7.5 IQ Signal Evaluation using Deep Learning

The adaptive algorithm presented in the Section 7.4 uses adaptive bandpass filtering to extract the heart rate signal along with a Kalman filter-based tracking procedure. The combination of these steps leads to narrowing of the passband range while simultaneously approximating the actual heart rate as the center frequency. Furthermore, the processing pipeline checks for deviation of the IQ data from circularity, and if the deviation is found to be large, the time segment is ignored since it implies data corrupted by random body movement and/or motion. The threshold needs to be set empirically and is always susceptible to the sensed human target. Thus, the algorithm step of comparing the standard deviation to a threshold needs to be enhanced to better classify what high-quality IQ data is. As well, this simple algorithm is incapable of distinguishing IMP in particular or harmonic corrupted data from HQD. In that context, in this section, this signal evaluation is proposed to be done through a 1D CNN. Figures 7.17, 7.18, and 7.19 presents exemplary IQ plots of HQD, data corrupted by RBM, and breathing-induced IMP/harmonics, respectively. The HQD is defined as one that eventually leads to correct estimation of the heart rate through the time-domain peak detection algorithm used in an earlier Section 7.4, while bad data are the ones that are corrupted data due to motion or RBM, and harmonics or IMP, leading to incorrect estimates through a peak detection algorithm.

The quality check of the target data is performed as a binary classification, wherein in one experiment the target is attempted to be classified as HQD as a class with IMP/harmonics, and RBM/motion data as another class, whereas in another

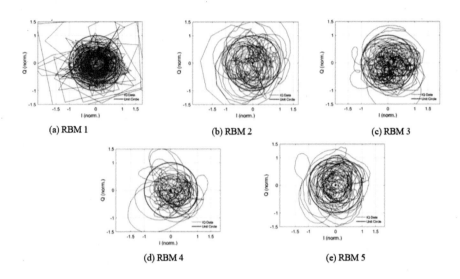

Figure 7.18   Exemplary IQ plots due to random body movement of the human target.

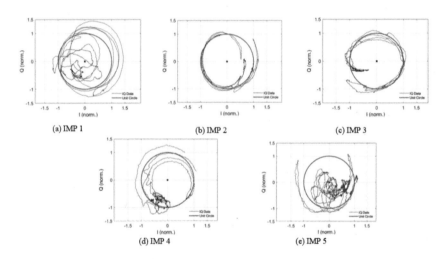

Figure 7.19   Exemplary IQ plots of IMP, IQ examples due to perturbation of segments due to intermodulation between breathing and heart rate and higher-order harmonics due to breathing.

experiment the target is attempted to be classified as HQD with IMP/harmonics as one class and RBM/motion data as another class, as presented in Figure 7.20. The ideal setting warrants the first experiment, where only the HQD time segment is allowed to update the tracked heart rate, but distinguishing between HQD and IMP/harmonics data can be difficult due to similar IQ visualizations. Target data is labeled in the training as IMP if the difference between estimated and reference

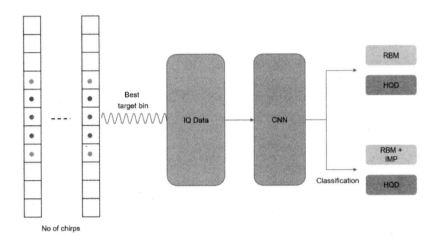

Figure 7.20 Proposed deep learning model.

heart rate is similar to the breath rate. For RBM the subject is asked to make small random movements during the recordings. Both the training and test data set contain time segments from 2,048 consecutive chirps, which are processed and fed into the deep neural network, as presented in Figure 7.20.

7.5.1 Deep Learning Architecture

The proposed architecture is based on 1D-CNN, consists of seven convolution layers, and takes an input of 256×1 one-dimensional signals. Each convolution layer uses a ReLU as the activation function and is followed by an average pooling layer of size 2×2 with strides 2 and batch normalization. The first convolution layer is composed of 1,024 convolution filters with a size of 3×3 and the second to seventh convolution layers use 512, 512, 128, 128, 128, and 64 convolution filters with a size of 3×3, respectively. After the second, third, and sixth convolution layer a dropout of 20% is applied. The last layer of the network is a fully connected layer, which uses softmax as an activation function.

An Adam optimizer is used for training the CNN. The CNN computes the adaptive learning rate for each network weight over the learning process from estimates of first and second moments of the gradients. In configuration parameters, the learning rate is set to $\alpha = 0.0002$ and the exponential decay rate for the first and second moment estimates are set to $\beta_1 = 0.5$ and $\beta_2 = 0.999$. The parameter that counters divide by zero problems is set to $\epsilon = 10^{-8}$.

7.5.2 Results and Discussion

A total of 3,110 measurement rows were obtained. Out of these, 2,610 measurement rows were used for training and 500 were used for testing. The

Table 7.4
Proposed System Performance

Methods	Class 1	Class 2	Accuracy	F_1 Score
Signal processing approach	HQD + IMP	RBM	63%	0.74
	HQD	RBM + IMP	43.2%	0.53
Deep learning method	HQD + IMP	RBM	99.8%	0.99
	HQD	RBM + IMP	85.4%	0.86

training data contains 45.70% of HQD signals, 32.98% of RBM, and 21.32% of IMP signals, respectively. The test data contains 44.80% of HQD signals, 32% of RBM, and 23.20% of IMP signals. The training data is divided into three classes. The first class contains the data corresponding to high-quality signals and is labeled as HQD, the second class represents IMP signals, and the third class are signals afflicted with RBMs.

The performance of the proposed system is evaluated by accuracy and F_1 score as shown in Table 7.4. Two classification settings are performed. In the first setting, RBM afflicted signals are classified against HQD and IMP signals, which results in an accuracy of 99.8% and an F_1 score of 0.99. The second setting separates HQD from RMBs as well as IMPs, whereby an accuracy of 85.4% and an F_1 score of 0.86 is achieved. For both the variants, the proposed method outperforms the signal processing approach based on estimating the circularity of the IQ data and comparing it to a threshold. The signal processing approach achieves 43.2% of accuracy with F_1 score of 0.53 for the first variant and achieves 63% of accuracy with F_1 score of 0.74 for the second variant. Thus, the deep learning based solution outperforms the classification/segregation capability offered by the signal processing approach. Also as expected, the classification performance in the case of first setting is better than from the second setting, since the IQ data for IMP and HQD are similar except for some artifacts due to second-order transfomation effects that result in IMP or harmonics.

7.6 Future Work and Direction

The presented hybrid solution, involving a deep learning approach and Kalman filter vital tracking, offers a robust, reliable, and accurate remote heart rate estimation using a single radar system for a single person in its field of view. However, the system has room for improvement to have accurate estimations under arbitrary conditions like aspect angle, sensor orientation, occlusions, and interfering Doppler sources, such as a fan or ventilator, in the field of view. Furthermore, there are algorithm changes required for the system to work for vital sensing of multiple people simultaneously since the nonideal sidelobe suppression window functions along fast-time would interfere with the phase signal of an adjacent human's range. As well system performance can be improved further through receiver beam-forming techniques to increase the signal-to-interference noise (SINR) ratio, which will improve the heart rate estimation in the presence of interference and sensing in the farther distance.

7.7 Problems

1. In the radar interferometric technique for heart rate estimation, what is the importance of phase estimation or angle calculation?
2. What are the filter cutoff frequencies required for a breathing signal and a heart signal?
3. The estimation of a heart rate signal is perturbed by intermodulation product and breathing harmonics. Justify.
4. How does updating the bandpass filter for heart rate estimation help in better heart rate estimation?
5. What does the gating operation in Kalman filter heart rate estimation achieve?
6. What happens to the Kalman filter–adaptive filter loop when the heart rate estimation for the consecutive next few frames is outside the bandpass filter set by the Kalman filter?
7. What is the need for GAN augmented data augmentation? What are the probable reasons GAN-based signal augmentation didn't result in good estimates through DeepNet?

References

[1] Gardner, R. M., and K. W. Hollingsworth, "Optimizing the Electrocardiogram and Pressure Monitoring," *Critical Care Medicine*, Vol. 14, No. 7, 1986, pp. 651–658.

[2] Li, C., Z. Peng, T.Y. Huang, et al., "A Review on Recent Progress of Portable Short-Range Noncontact Microwave Radar Systems," *IEEE Transactions on Microwave Theory and Techniques*, Vol. 65, No. 5, 2017, pp. 1692–1706.

[3] Li, C., V. M. Lubecke, O. Boric-Lubecke, and J. Lin, "A Review on Recent Advances in Doppler Radar Sensors for Noncontact Healthcare Monitoring," *IEEE Transactions on Microwave Theory and Techniques*, Vol. 61, No. 5, 2013, pp. 2046–2060.

[4] Li, Changzhi, J. Lin, and Y. Xiao, "Robust Overnight Monitoring of Human Vital Signs by a Non-Contact Respiration and Heartbeat Detector," 2006 International Conference of the IEEE Engineering in Medicine and Biology Society, 2006, pp. 2235–2238.

[5] Li, J., L. Liu, Z. Zeng, and F. Liu, "Advanced Signal Processing for Vital Sign Extraction with Applications in UWB Radar Detection of Trapped Victims in Complex Environments," *IEEE Journal of Selected Topics in Applied Earth Observations and Remote Sensing*, Vol. 7, No. 3, 2013, pp. 783–791.

[6]   Gu, C., and C. Li, "From Tumor Targeting to Speech Monitoring: Accurate Respiratory Monitoring Using Medical Continuous-Wave Radar Sensors," *IEEE Microwave Magazine*, Vol. 15, No. 4, 2014, pp. 66–76.

[7]   Vinci, G., T. Lenhard, C. Will, and A. Koelpin, "Microwave Interferometer Radar-Based Vital Sign Detection for Driver Monitoring System," in *2015 IEEE MTT-S International Conference on Microwaves for Intelligent Mobility (ICMIM)*, April 2015, pp. 1–4.

[8]   Gu, C., "Short-Range Noncontact Sensors for Healthcare and Other Emerging Applications: A Review," *Sensors*, Vol. 16, No. 8, 2016, p. 1169.

[9]   Van Loon, K., M. J. M. Breteler, L. Van Wolfwinkel, et al., 2016. "Wireless Non-Invasive Continuous Respiratory Monitoring with FMCW Radar: A Clinical Validation Study," *Journal of Clinical Monitoring and Computing*, Vol. 30, No. 6, pp. 797–805.

[10] Santra, A., R.V. Ulaganathan, and T. Finke, "Short-Range Millimetric-Wave Radar System for Occupancy Sensing Application," *IEEE Sensors Letters*, Vol. 2, No. 3, 2018, pp. 1–4.

[11] C. Will, K. Shi, S. Schellenberger, et al., 2017, "Local Pulse Wave Detection Using Continuous Wave Radar Systems," *IEEE Journal of Electromagnetics, RF and Microwaves in Medicine and Biology*, Vol. 1, No. 2, pp. 81–89.

[12] Odinaka, I., J. A. O'Sullivan, E. J. Sirevaag, and J. W. Rohrbaugh, "Cardiovascular Biometrics: Combining Mechanical and Electrical Signals," *IEEE Transactions on Information Forensics and Security*, Vol. 10, No. 1, 2014, pp. 16–27.

[13] Tu, J., and J. Lin, "Fast Acquisition of Heart Rate in Noncontact Vital Sign Radar Measurement Using Time-Window-Variation Technique," *IEEE Transactions on Instrumentation and Measurement*, Vol. 65, No. 1, 2015, pp. 112–122.

[14] Vinci, G., S. Lindner, and F. Barbon, et al., "Six-Port Radar Sensor for Remote Respiration Rate and Heartbeat Vital-Sign Monitoring," *IEEE Transactions on Microwave Theory and Techniques*, Vol. 61, No. 5, 2013, pp. 2093–2100.

[15] Immoreev, I. Y., "Ultrawide band Radars: Features and Capabilities," *Journal of Communications Technology and Electronics*, Vol. 54, No. 1, 2009, pp. 1–26.

[16] Mostov, K., E. Liptsen, and R. Boutchko, "Medical Applications of Shortwave FM Radar: Remote Monitoring of Cardiac and Respiratory Motion," *Medical Physics*, Vol. 37, No. 3, 2010, pp. 1332–1338.

[17] Xiong, Y., S. Chen, X. Dong, Z. Peng, and W. Zhang, "Accurate Measurement in Doppler Radar Vital Sign Detection Based on Parameterized Demodulation," *IEEE Transactions on Microwave Theory and Techniques*, Vol. 65, No. 11, 2017, pp. 4483–4492.

[18] Chioukh, L., H. Boutayeb, D. Deslandes, and K. Wu, "Noise and Sensitivity of Harmonic Radar Architecture for Remote Sensing and Detection of Vital Signs," *IEEE Transactions on Microwave Theory and Techniques*, Vol. 62, No. 9, 2014, pp. 1847–1855.

[19] Santra, A., R. V. Ulaganathan, T. Finke, et al., "Short-Range Multi-Mode Continuous-Wave Radar for Vital Sign Measurement and Imaging, in *2018 IEEE Radar Conference (RadarConf18)*, April 2018, pp. 0946–0950.

[20] Sakamoto, T., R. Imasaka, H. Taki, et al., "Feature-Based Correlation and Topological Similarity for Interbeat Interval Estimation Using Ultrawide band Radar," *IEEE Transactions on Biomedical Engineering*, Vol. 63, No. 4, 2015. pp. 747–757.

[21] Hu, W., Z. Zhao, Y. Wang, H. Zhang, and F. Lin, "Noncontact Accurate Measurement of Cardiopulmonary Activity Using a Compact Quadrature Doppler Radar Sensor," *IEEE Transactions on Biomedical Engineering*, Vol. 61, No. 3, 2013, pp. 725–735.

[22] Massagram, W., V. M. Lubecke, A. Hst-Madsen, and O. Boric-Lubecke, "Assessment of Heart Rate Variability and Respiratory Sinus Arrhythmia via Doppler Radar," *IEEE Transactions on Microwave Theory and Techniques*, Vol. 57, No. 10, 2009, pp. 2542–2549.

[23] Will, C., K. Shi, F. Lurz, R. Weigel, and A. Koelpin, "Instantaneous Heartbeat Detection Using a Cross-Correlation Based Template Matching for Continuous Wave Radar Systems," *IEEE Topical Conference on Wireless Sensors and Sensor Networks (WiSNet)*, Austin, TX, 2016, pp. 31–34.

[24] Will, C., K. Shi, R. Weigel, and A. Koelpin, "Advanced Template Matching Algorithm for Instantaneous Heartbeat Detection Using Continuous Wave Radar Systems," in *2017 First IEEE MTT-S International Microwave Bio Conference (IMBIOC)*, May 2017, pp. 1–4.

[25] Lv, Q., Y. Dong, Y. Sun, C. Li, and L. Ran, "Detection of Bio-Signals from Body Movement Based on High-Dynamic-Range Doppler Radar Sensor," in *2015 IEEE MTT-S 2015 International Microwave Workshop Series on RF and Wireless Technologies for Biomedical and Healthcare Applications (IMWS-BIO)*, September 2015, pp. 88–89.

[26] Tu, J., T. Hwang, and J. Lin, "Respiration Rate Measurement Under 1-D Body Motion Using Single Continuous-Wave Doppler Radar Vital Sign Detection System," *IEEE Transactions on Microwave Theory and Techniques*, Vol. 64, No. 6, 2016, pp. 1937–1946.

[27] Gu, C., J. Wang, and J. Lien, "Deep Neural Network Based Body Movement Cancellation for Doppler Radar Vital Sign Detection," in *2019 IEEE MTT-S International Wireless Symposium (IWS)* May 2019, pp. 1–3.

[28] Will, C., K. Shi, S. Schellenberger, et al., "Radar-Based Heart Sound Detection," *Scientific Reports*, Vol. 8, No. 1, 2018, pp. 1–14.

[29] Chin, T.Y., K. Y. Lin, S. F. Chang, and C. C., Chang, "A Fast Clutter Cancellation Method in Quadrature Doppler Radar for Noncontact Vital Signal Detection," in *2010 IEEE MTT-S International Microwave Symposium*, May 2010, pp. 764–767.

[30] Kuo, H. C., H. H. Wang, P. C. Wang, H. R. Chuang, and F. L. Lin, "60-GHz Millimeter-Wave Life Detection System with Clutter Canceller for Remote Human Vital-Signal Sensing," in *2011 IEEE MTT-S International Microwave Workshop Series on Millimeter Wave Integration Technologies*, September 2011, pp. 93–96.

[31] Yang, S., H. Qin, X. Liang, and T. A. Gulliver, "Clutter Elimination and Harmonic Suppression of Non-Stationary Life Signs for Long-Range and Through-Wall Human Subject Detection Using Spectral Kurtosis Analysis (SKA)-Based Windowed Fourier Transform (WFT) Method," *Applied Sciences*, Vol. 9, No. 2, 2019, p. 355.

[32] Liang, X., H. Zhang, T. Lu, H. Xiao, G. Fang, and T. A. Gulliver, "An Improved Signal Processing Algorithm for VSF Extraction," *Multidimensional Systems and Signal Processing*, Vol. 30, No. 4, 2019, pp. 1811–1827.

[33] Li, C., and J. Lin, "Random Body Movement Cancellation in Doppler Radar Vital Sign Detection," *IEEE Transactions on Microwave Theory and Techniques*, Vol. 56, No. 12, 2008, pp. 3143–3152.

[34]   Gu, C., G. Wang, T. Inoue, and C. Li, "Doppler Radar Vital Sign Detection with Random Body Movement Cancellation Based On Adaptive Phase Compensation," in *2013 IEEE MTT-S International Microwave Symposium Digest (MTT)*, June 2013, pp. 1–3.

[35]   Wang, F. K., T. S. Horng, K. C. Peng, J. K. Jau, J. Y. Li, and C. C. Chen, "Single-Antenna Doppler Radars Using Self and Mutual Injection Locking for Vital Sign Detection with Random Body Movement Cancellation," *IEEE Transactions on Microwave Theory and Techniques*, Vol. 59, No. 12, 2011, pp. 3577–3587.

[36]   Mostafanezhad, I., E. Yavari, O. Boric-Lubecke, V. M. Lubecke, and D. P. Mandic, "Cancellation of Unwanted Doppler Radar Sensor Motion Using Empirical Mode Decomposition," *IEEE Sensors Journal*, Vol. 13, No. 5, 2013, pp. 1897–1904.

[37]   Mercuri, M., I. R. Lorato, Y. H. Liu, F. Wieringa, C. Van Hoof, and T. Torfs, "Vital-Sign Monitoring and Spatial Tracking of Multiple People Using a Contactless Radar-Based Sensor," *Nature Electronics*, Vol. 2, No. 6, 2019, pp. 252–262.

[38]   Levick, J. R., *An Introduction to Cardiovascular Physiology*, London: Taylor & Francis, 2003, pp. 1–13.

[39]   Peng, Z., and C. Li, "Portable Microwave Radar Systems for Short-Range Localization and Life Tracking: A Review," *Sensors*, Vol. 19, No. 5, 2019, p. 1136.

[40]   Santra, A., I. Nasr, and J. Kim, "Reinventing Radar: The power of 4D Sensing," *Microw. J.*, Vol. 61, No. 12, 2018, pp. 26–38.

[41]   Singh, A., Gao, X., Yavari, E., et al., "Data-Based Quadrature Imbalance Compensation for a CW Doppler Radar System," *IEEE Transactions on Microwave Theory and Techniques*, Vol. 61, No. 4, 2013, pp. 1718–1724.

[42]   Arsalan, M., A. Santra, and C. Will, "Improved Contactless Heartbeat Estimation in FMCW Radarvia Kalman Filter Tracking," *IEEE Sensors Letters*, Vol. 4, No. 5, May 2020, pp. 1–4.

8

People Sensing, Counting, and Localization

8.1 Introduction

In the United States, buildings are responsible for consumption of 42% of total energy. In residential and commercial areas the energy consumption for lighting and HVAC systems are about 30% and over 50% [1,2], respectively. Several studies have shown that accurate knowledge of occupancy and people count in residential, commercial, or public spaces can save energy consumption by 25%–75% [3] through implementation of a smart feedback loop to control lighting and HVAC systems automatically [4]. However, due to privacy concerns, implementing systems that is cognizant of their occupants through visual sensors is not possible. This demands the use of alternate and efficient occupancy sensors, people-counting sensor systems, and people localization systems, which can enable intelligent and effective use of energy resources.

Passive infrared sensors (PIRs) and ultrasonic sensor based solutions are solutions available in the market for presence/occupancy sensing. However, the major drawback of such sensors are that they are prone to false alarms, leading to ineffective energy regulation. The main shortcoming is that they can detect major body movements, such as walking or large arm movements, but when the person is quasi-static, such as working on a computer or watching TV or sleeping, the sensor generates false negatives. Additionally, the sensitivity of the PIR sensors drops off drastically with distance and requires a line of sight other than being prone to interference [5,6]. Furthermore, PIR systems are temperature-dependent, which can result in vast numbers of false detections arising from environmental variations. On the other hand, sound waves from ultrasonic sensors typically are absorbed or reflected by different cloth or foam materials and have high sensitivity to reflective materials such as glass or plastic, leading to false alarms. Although ultrasound sensors do not require line of sight and work well for longer distances, they are prone to false positives due to environmental factors such as air turbulences.

Radar has evolved from automotive applications such as driving assistance systems, safety and driver alert systems, and autonomous driving systems to low-cost solution penetrating industrial and consumer market segments. Radar has been used for perimeter intrusion detection systems [7], gesture recognition [8,9], human-machine interfaces [10], outdoor positioning and localization [11], and indoor people counting [12]. The significant challenge for wide adoption of radar sensors for occupancy sensing is to demonstrate reliable system performance at low power and low cost. Current occupancy sensors are significantly limited since they only serve as motion sensors and are not truly presence/occupancy sensors. The ability of radars to remotely sense and detect minute motion due to human cardiopulmonary and breathing makes them a promising solution to address the problems of false triggers and dead spots in conventional sensors [13–15]. Several Doppler based radar sensors for occupancy sensing have been proposed in literature [16–19].

Human sensing in general can be classified into several categories based on the information that needs to be extracted from the operating environment. In terms of information, the lowest level information of cognizance is occupancy/presence (i.e., if there is no one or at least one person present in the operating environment). Presence sensors, often called binary sensors, are the most-sought-after application and also the simplest in terms of processing and computation requirements. Presence sensing enables lights to turn on/off or computers to log on/off based on the binary decision whether no one or someone is present. With some level of signal processing utilizing adaptive thresholding and detecting vital signals, radars and RF sensors can offer reliable presence/occupancy sensing. However, radar sensors enabled through signal processing are still not immune to pets or false alarms due to fans, ventilators, moving curtains, or plants. Such issues can be resolved through deep learning based approaches, making short-range radars truly presence/occupancy sensors.

The next level of information of cognizance that can be extracted is the people count, which essentially counts how many people are in the operating environment. People counting in an indoor or outdoor environment has lot of applications. People counting in indoor rooms can assist in air flow control strategies of HVAC systems and smart lighting systems, by regulating the operating mode [20]. People counting system can also help administrations in automatic generation of footfall statistics in malls or transportation systems, which can be subsequently used for determining conversion rates and better planning of services. Furthermore, people counting systems find applications in elevators or areas where limitation on the usage by a maximum number of people at any time is mandated. Several people counting sensors based on cameras, thermal imaging, infrared beams, and radar sensors have been proposed and are available on the market. Although camera-based people counting systems using deep learning have proven to be reliable and accurate in counting people in dense crowds or clutter environments [21–26], they inherently suffer from privacy concerns and are prone to illumination conditions, making them unfavorable for industrial and outdoor settings. Although radar-based people counting solutions [27–32] offer immunity to such issues, radar-based human detection poses several challenges, such as missed detection due to shadowing or occlusion, poor resolution data for target identifiability, and time-varying radar signal strength arising due to the super position of reflections from multiple body parts. Owing to these challenges, human detection under dense environments using low-cost, short-range radars with conventional approaches is a difficult and intractable problem. Several of these challenges can be overcome through deep learning approaches, which we discuss later in the chapter.

The third level of information is people localization in the operating environment (i.e., detecting and specifically locating where the people are in the field of view). Localizing people in the indoor environment can help in energy preservation through directed or customized service; for example once the location of people in the room are detected an air conditioner can be directed either toward the location where people are or to other locations where people are not present. Further, in an outdoor surveillance environment where radar is sensor-fused with a camera, radar target detection and localization can be used to direct the camera to zoom into the location/area of interest. Thus, human localization in both indoor and outdoor environments has several application use-cases. The information can be extracted through a camera or ToF or sensors, an array of ultrasound sensors or radar sensors. FMCW radars can provide a ubiquitous solution to sense and localize human targets. Human target detection in indoor environments poses further challenges, such as ghost targets from static targets such as walls, chairs, and furniture, and also spurious radar responses due to multipath reflections from multiple targets [33]. Further, often a strong reflecting or closer human targets occlude less reflecting or farther human targets at the detector output. While the earlier phenomenon leads to over-estimating the count of humans, the latter phenomenon leads to under-estimating the count of humans in the room. This results in increased false alarms and low detection probabilities, leading to poor radar receiver operating characteristics and inaccurate application-specific decisions based on human target counts or target tracking. In this chapter, we demonstrate how deep learning approaches can be used to improve human target detection/localization in such challenging environments.

The next level of information is recognizing the human activity—if a person is walking, exercising, sitting, or sleeping. Based on the human activity, the HVAC

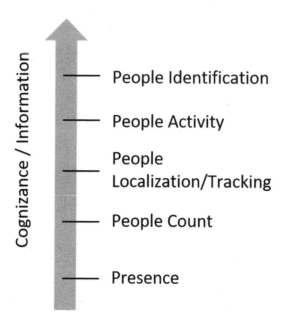

Figure 8.1   Information/cognizance level of human sensing.

or lighting can automatically change to low, medium, or high mode, facilitating intelligent energy savings. Typically, sensing human activity follows after human detection and tracking the person spatiotemporally (i.e., what the person was doing and where at an earlier time instance). An even higher level of information extraction is human identification. Such information can be sensed for customized and preferred usage or settings, and also enable applications such as automatic authentication and recognition. In this chapter, we do not focus on these two aspects of information, and concentrate on how deep learning assists in lower-level of information components. We present the different components of information complexity pertaining to various levels of human sensing in Figure 8.1.

In this chapter, we discuss novel signal processing architecture based on a finite state machine to achieve presence/occupancy sensing using short-range radars. We discuss how different system parameters are chosen to meet system requirements. We present the limitations of the signal processing approach and how human classification through deep learning approach offers a robust and reliable solution for a ubiquitous presence/occupancy solution.

Further, in this chapter, we present novel approach to count people that is based on processing architecture that extracts macro-Doppler, micro-Doppler, and vital-Doppler motions and process them seperately to deal with challenges associated with differing properties of these Doppler components and thus offering a robust signal processing pipeline for people counting. However, this radar-based signal processing approach also suffers from limitations and frequent incorrect counts. We propose two deep learning solutions, one based on N-pair loss using meta-learning ideas, and another based on cross-learning using supervised information from the camera. A novel multimodal cross-learning framework to distill high-level feature abstraction learned from camera-based DCNN to radar DCNN is proposed, so that during testing/inference the radar DCNN alone can deliver a far superior performance compared to its counterpart radar-only DCNN in counting people.

Subsequently, we present the topic of human localization. Radar responses from human targets are in general spread across Doppler due to the macro-Doppler component due to torso and associated micro-Doppler components due to hand, shoulder, and leg movements. With the use of higher sweep bandwidths, the radar echoes from targets are not received as point targets across range but are spread across the range and are referred to as range-extended targets. Thus, human targets are perceived as doubly spread targets [34] across range and Doppler. Thereby, human target detections using radars with higher sweep bandwidths require several adaptations in the standard signal processing pipeline before feeding into application-specific processing (e.g., target counting or target tracking). Conventionally, the radar range-Doppler image (RDI) processing pipeline involves MTI to remove static targets, and maximal ratio combining (MRC) to integrate data across antennas. The RDI is then fed into a constant-false alarm rate (2D-CFAR) detection algorithm, which then detects whether a cell under test (CUT) is a valid target or not by evaluating the statistics of the reference cells. The constant false alarm rate detection adaptively calculates the noise floor by calculating the statistics around the CUT and guarantees a fixed false alarm rate, which sets the threshold multiplier with which the estimated noise floor is scaled. Further, the detections are grouped together using a clustering algorithm so that reflections from the same human target are detected as single cluster and likewise for different humans. But the standard signal processing approach suffers from several drawbacks in terms of ghost targets and missed detections, and therefore propose to overcome such limitations through a deep learning based processing pipeline. A deep residual U-Net is proposed to process raw RDIs to generate target detected RDIs. The target detected RDIs are demonstrated to suppress reflections from ghost targets, reject spurious targets due to multipath reflections, avoid target occlusions, and achieve accurate target clustering in the detected RDI, thus resulting in reliable and accurate human target detections.

The chapter is laid out as follows: in Section 8.2, we present a human presence sensing solution using a signal processing approach after outlining the general system requirements and challenges of presence sensing solution. In Section 8.3, we present the challenges and drawbacks of a signal processing-based radar presence solution and we present a deep learning based solution to overcome the challenges. We present the application of a people counting through signal processing approach in Section 8.4 and deep learning approach in Section 8.5. In Section 8.5, we present a novel cross-learning based solution, where supervised information from synchronized camera is used for improved training of a radar-based count classifier. In Sections 8.6 and 8.7, we present a signal processing approach and deep learning approach to people detection and localization. In Section 8.7, we present and compare the detection performance of a signal processing and deep learning solution. We conclude with future work and research directions in Section 8.8.

8.2 Presence Sensing: Signal Processing Approach

8.2.1 Challenges

The presence sensing solution is implemented through a finite state machine, wherein one of the state is in an absence state and the other state is in a presence state. The motivation to maintain two states are majorly driven by system requirements mainly; for example, the power consumption during the absence state should be minimum and should consume less power than the idle state of the device on which the radar is mounted. Thus the modulation scheme and the number of chirps per second in the absence state is kept minimum with a single chirp per frame and the frame being 512 ms apart. The transmit power is also kept low to minimize power consumption. The presence state has a much higher number of chirps per frame, with the chirp spacing of 1 ms and having 512 chirps per frame. Figure 8.2 presents the two state machines along with different parameters and settings, such as active zone, adaptive threshold factor, modulation scheme, and transmit power in both the states. Figure 8.3 depicts an exemplary

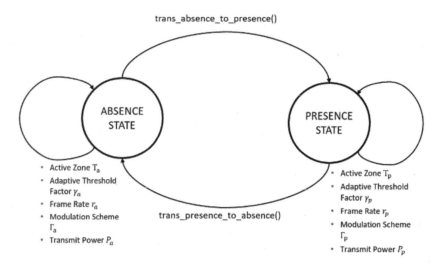

Figure 8.2 State machine operating in a absence state and a presence state for presence sensing.

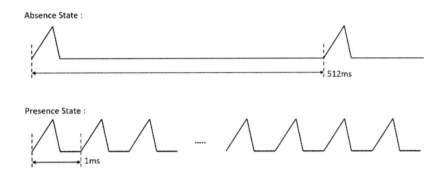

Figure 8.3 Chirp modulation scheme during presence and absence states.

modulation scheme involving the number of chirps and pulse repetition time, in both the states.

The motivation to maintain two different states are conflicting requirements during both phases. The more stringent constraint during the absence state is to prevent or avoid false positives; that is, the system shouldn't detect that a target is present when none is. While in the presence state the more stringent requirement is to minimize true negatives; that is, the person is present and the system shouldn't detect absence. The two conflicting requirements can be satisfied by maintaining the state machine and assigning different adaptive threshold factors during presence (lower threshold) and absence (higher threshold) states. Thus, state machine implementation helps meet system requirements for power consumption and also improves the radar operating characteristics by minimizing false positives (probability of false alarm, Pfa) and true negatives (thus increasing probability of detection, Pd).

8.2.2 Solution

The signal processing operations performed in an absence or presence state is outlined below, including the conditions for transition from one state to another. Except for system parameters, the processing operations in both states are the same. Figure 8.4 presents the processing pipeline used during the absence and presence states.

1. Moving Target Indication Filtering
   MTI filters in principle are low-order, simple finite impulse response (FIR) designs. At each time stamp the absolute maximum value over the slow time of each range bin is denoted by $r_{i,\max}$. The MTI filter value t_i is the weighted average of this maximum value and the previous MTI

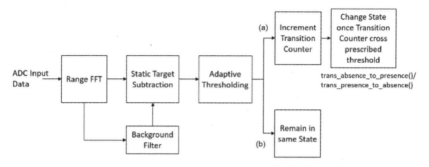

Figure 8.4 Data processing pipeline during both absence state and presence state. Path (a) is chosen when at least one target range bin is detected, and path (b) none is detected during absence state. Whereas path (a) is chosen when no target range bin is detected, path (b) is chosen when at least one target range bin is detected.

filter value t_{i-1} with a weight of α:

$$t_i = \alpha \cdot r_{i,\text{max}} + (1 - \alpha) \cdot t_{i-1}. \tag{8.1}$$

In the first time stamp t_0 is initialized with zero. For each range bin, the previous MTI filter value is subtracted from $r_{i,\text{max}}$ to obtain the filtered range FFT value $r_{i,\text{filt}}$:

$$r_{i,\text{filt}} = \text{abs}(r_{i,\text{max}} - t_{i-1}). \tag{8.2}$$

This filtered value is then utilized for the subsequent target detection. MTI processing performs a linear filtering that leads to a diminished signal strength for static targets while maintaining the signal strength of the moving targets. Therefore, MTI processing helps to remove completely static targets, while nonstatic targets like humans are retained.

2. Coherent Integration

Optionally, coherent integration is performed to increase the signal-to-noise ratio by combining the phase and magnitude of the range FFT data coherently over subsequent chirps. Coherent integration is based on the mean R_{mean} of the range FFT data r_i over N_C chirps:

$$R_{\text{mean}} = \frac{1}{N_C} \cdot \sum_{i=1}^{N_C} r_i \tag{8.3}$$

$$r_i = a_i \cdot e^{j\phi_i} \tag{8.4}$$

Here, amplitude and phase of the range FFT value for chirp i are denoted by a_i and ϕ_i, respectively. The coherent integration is optionally performed over consecutive chirps across multiple frames.

3. Adaptive Thresholding

   After the MTI filter and optional coherent integration of consecutive chirps, the L2 norm of the range data is calculated. The L2 norm represents the average noise floor across all range bins, which is then multiplied with the threshold factor. In the absence state, this threshold factor is kept high (e.g., 2.5-3) to avoid false positives. In the presence state, this threshold factor is kept low (e.g., 1.5-2) to avoid true negatives. This adaptively computed value is for thresholding the current range data, and if any range bin within the active zone crosses this threshold, a human target is declared to be detected. However, if none of the range bins cross this computed threshold, no human target is declared to be detected.

4. Remain in the Same State

   In the case of a presence state, if the current frame or computation has detected a target, the presence state is continued. However, if the current frame or computation has no detections, the absence state is continued.

5. Increment Counter and Change State

   In the case of the presence state, if the current frame or computation has no target detected a global counter is incremented, which by default is set to 0. If the subsequent frames or computations also have no target, the counter is accordingly incremented until it reaches a prescribed value, following which the FSM changes state to absence by changing the appropriate system parameters through trans_presence_to_absence( ) application programming interface (API) to change hardware parameters to transition from presence to absence state. The global counter is reset back to 0 after the transition or if a target is detected on subsequent frames. Similarly in the case of an absence state, if the current frame or computation has a target detected, the global counter is incremented. It is consecutively incremented on detection of target until the global counter reaches a prescribed value, following which the FSM changes state to presence. Accordingly, the system parameters for the presence state are set through trans_absence_to_presence( ) API. The global counter acts as hystersis to avoid spurious toggling between the presence and absence states, and further, the purpose is to counter ghost targets in the absence state and missed detections in the presence state.

8.3 Presence Sensing: Deep Learning Approach

8.3.1 Challenges

Although the presence sensing solution presented in the previous section works quite well in detecting moving targets and their absence, it fails in recognizing human targets from other moving or vibrating targets. Figure 8.5 presents the

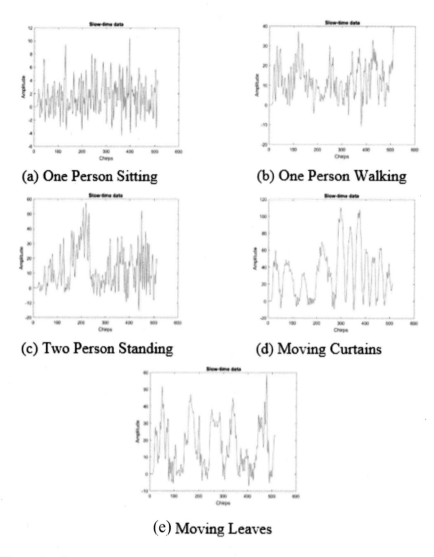

Figure 8.5 Exemplary extracted slow-time data for (a) one person sitting with minor movements, (b) one person walking around in the room, (c) two people standing, (d) moving curtains due to wind, and (e) moving plant leaves due to wind.

extracted slow-time data for one of the frame example when a person sitting is in a chair and making minor movements (a), one person walking around in the field of view (b), two people standing (c), moving curtains due to a mildly blowing wind (d), and moving plant leaves due to wind (e). This slow-time data of various moving/vibrating targets in indoor environments suggests that the signal processing-based presence sensing solution would get triggered into the presence state due to non-human vibrating target sources.

In an indoor environment, other vibrating sources such as vibrating window panes or ventilators can spuriously trigger the sensor into the presence state. For a truly ubiquotous and reliable solution, it is important to recognize human targets even in the presence of other vibrating/moving targets. This can be achieved through deep learning based solution. We present a SincNet-based solution using focal binary entropy loss function to classify between human vs. everything else.

8.3.2 Solution

Figure 8.6 presents the modified processing pipeline in the presence state to enable classification of human vs. everything else. In the modified pipeline, following target detection in the fast-time after MTI filter, the slow-time data is extracted from the detected range bin across the frame. Based on the human or non human classification, the presence state is accordingly activated or moved back to the absence state.

In the typical CNN architecture, the first layers perform convolution between a bank of 2D kernels and the input image map. In the standard radar signal processing approach, the first step requires an spectral analysis along the slow time followed by detection and classification of the data. Alternately, to process 1D signals such as radar, the slow-time data preprocessing step often involves convolution with a finite impulse response filter

$$y_{out}[n] = y_{in}[n] * h[n] \qquad (8.5)$$

where $*$ denotes the convolution operation, and $y_{in}[n]$ being the input signal. In a standard 1D convolution neural network, $h[n]$ would be learned by starting

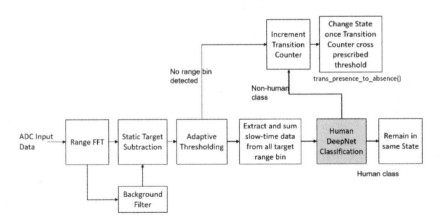

Figure 8.6 Modified processing pipeline in the presence state with incorporation of presence deepNet.

from a random set of values. A reasonable filtering approach in the case of radar slow-time data and other signal processing operations, is sinc() function operating in different frequency bands. This is inspired from a generic bank of filters and wavelet transforms. Thus, instead of learning arbitrary convolution filters $h[n]$, the learning is restricted to parametrized learning structures $h[n; \theta]$, in the case of SincNet $\theta = [f_1, f_2]$. Thus the convolutional filter structure in SincNet [45] is described as

$$h[n; f_{start}, f_{end}] = f_{end} sinc(2\pi f_{end} n) - f_{start} sinc(2\pi f_{start} n) \qquad (8.6)$$

where $sinc(x) = sin(x)/x$. The bank of 1D-convolution filters $h^k[n; f_{start}, f_{end}]$ where $k \in (1, \ldots, K)$.

However, a traditional bandpass filter with sharp start and end frequencies causes ripples and high sidelobes in other adjacent frequencies. A common approach to avoid such sidelobes and ripples is to multiply the time-domain filter response with window functions $w[n]$ such as a Hamming window, thus the parameterized kth kernel is expressed as

$$\hat{h}^k[n; f_{start}, f_{end}] = h^k[n; f_{start}, f_{end}].w[n] \qquad (8.7)$$

During the training phase, the parameters (start and end frequencies) of each filter kernel are learned. The learned parameters of each kernel are optimized for the human vs. all classification tasks.

8.3.2.1 Architecture, Data, and Learning

Figure 8.7 presents the SincNet-based architecture to classify human against nonhuman. The slow-time radar data is extracted from chirps 1 ms apart, and thus the sampling frequency is 1 kHz. In a frame, 512 chirps are collected resulting in the observation window of 0.512s, which is fed into Sinc-Net architecture.

The parameterized start and end frequencies of each kernel are learned between 0 Hz to 1 kHz. The first layer performs parameterized Sinc-based convolutions using 50 filters of length L = 15 samples. Along the lines of the original SincNet proposed for speaker recognition, our proposed radar model for presence classification then employs two standard 1D convolutional layers, using 60 filters of length 7 each. A dropout of 0.5 and 1D max-pooling layer with a pool size of 3 are used after every convolution and a Sinc-based convolution layer to prevent overfitting and reduce the number of trainable parameters. Next, a fully connected layer composed of 128 neurons and dropout of 0.5 were applied. All layers use ReLU activation function. For initialization, all weights except that of Sinc-layers were initialized through Glorot initializer. The parameters of the Sinc-layer were initialized uniformly between 0 Hz to 1 kHz at steps of 20 Hz. The human presence classification is obtained by applying sigmoid activation at

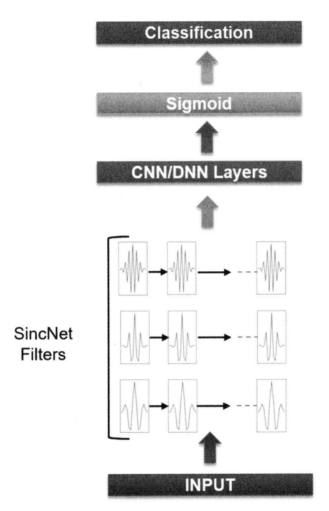

Figure 8.7   Human presence classification using SincNet discriminating against other moving targets.

the last layer. The network was trained using RMSprop optimizer with learning rate of 0.001, with a minibatch size of 64 samples.

The human presence classification data set has complex IQ slow-time data with each example being 512 samples long. The human data set included data such as a single static human, a person sitting idle, a person sitting and moving their arm, a person standing still, a person standing and exercising, and a person moving around. Additionally the human data set contains two people moving around, two people sitting and two people standing. Overall the human data set had 5,210 examples, and the other class had data of an empty room with furniture. Further the nonhuman class includes data from moving curtains, moving plants, and a vibrating ventillator, which are typically sources of false alarm using signal processing approaches. Before feeding into the radar Sinc-Net, only the real part of the complex IQ data is extracted and fed into the deep Net. The implication of this is that we lose the direction information of Doppler, whether the target is approaching or retreding. This however is not a problem for human presence classification.

8.3.2.2 Loss Function

For training the model, focal loss [43] on binary entropy was used as explained in the following equation;

$$FL(p_t) = \left(1 - p_t\right)^{\gamma} \log(p_t) \qquad (8.8)$$

The variables $y \in \{\pm 1\}$, $p \in [0, 1]$, and $p_t \in [0, 1]$, specify the class label, the estimated probability for the class with label $y = 1$, and the probability that a pixel was correctly classified, as defined by

$$p_t = \begin{cases} p & \text{if } y = 1 \\ 1 - p & \text{otherwise} \end{cases} \qquad (8.9)$$

The purpose of using focal loss is to deal with class imbalance in the training data due to the nature of the target setup, and collecting human data is easier compared to nonhuman example data. The focal loss places a higher weight on the binary-entropy loss for misclassified class and a much lower weight on well-classified ones. Thus, the rare class is in general better dealt with during training.

8.3.3   Results and Discussion

Figure 8.8 presents the time domain and frequency response of some learned Sinc-layer kernels. Figure 8.9 presents the effective frequency response of the learned SincNet kernels over the frequency space.

With the proposed radar SincNet-based model, we achieve an overall F1 score of 0.98 in being able to classify human vs. nonhuman, compared to an F1 score of 0.75 that is achieved if the standard convolutional layers are used. Apart from the fact that 1D standard convolution Net performs poorly in this classification problem mainly due to the absence of any filter structure in its convolution kernels, the additional advantage of using SincNet compared to comparable standard 1D convolution is that the number of parameters required are substantially less, making them a favorable option for embedded implementation.

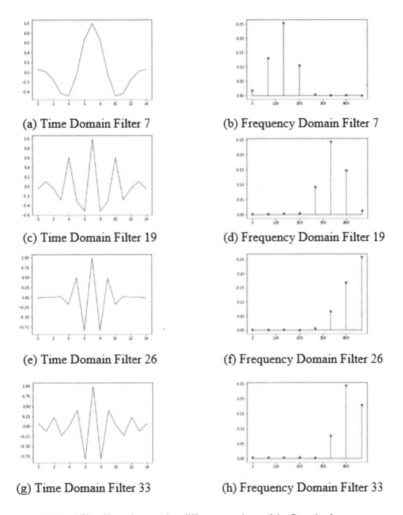

(a) Time Domain Filter 7

(b) Frequency Domain Filter 7

(c) Time Domain Filter 19

(d) Frequency Domain Filter 19

(e) Time Domain Filter 26

(f) Frequency Domain Filter 26

(g) Time Domain Filter 33

(h) Frequency Domain Filter 33

Figure 8.8 Learned Sinc filters for sensing different portions of the Doppler frequency spectrum.

8.4 People Counting: Signal Processing Approach

8.4.1 Challenges

People counting using radar is a challenging problem under practical scenarios when different activities, such as walking, running, sitting idle, and working on computer are to be considered. In general, human motion can be categorized into three different categories: namely macro-Doppler, micro-Doppler, and vital-Doppler. Large-amplitude or gross bodily movements of humans are referred to as macro-Doppler motions, such as running or walking. Small-amplitude bodily movements, bodily gestures, such as a person working on a computer, are referred as micro-Doppler motions. The pulmonary periodic motion due

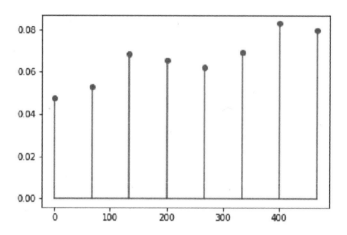

Figure 8.9   Effective frequency response of the learned SincNet kernels over the entire Doppler frequency band.

to cardiac or respiratory signal of humans sensed by the radar is referred as vital-Doppler motions. Radar returns from large-amplitude movements, such as walking, ascending, and descending stairs, typically have high Doppler frequency components, have large amplitude, have less spread in time-frequency plane (i.e., bursty). On the other hand, smaller-amplitude micro-Dopplers are due to movements corresponding to minor body movements such as watching TV, eating, cooking, or working on a computer have larger spread along time-frequency plane while having lower Doppler frequency components. The vital-Doppler motion operates in an even lower frequency band corresponding to the vital signal of the human body and has even lower amplitude. However, for people counting in a general setting would require adapting the sensor settings and parameters to sense the differing Doppler components. Figure 8.10 depicts the various human activities and the associated Doppler components that are detectable by the FMCW radar.

In general in FFT spectra, the high-amplitude Doppler frequency sways over the lower amplitude Doppler frequency due to its sidelobes in the same range bin, similar to how a high RCS target shadows a neighboring low RCS target in the range FFT. Due to the different properties of macro-Dopple, micro-Doppler motions, and vital-Doppler, using the same range-Doppler map is not effective for people counting under all practical human activities. Setting a higher threshold would be able to detect large macro-Doppler motions like walking and running, but micro-Doppler motions like working on computer or watching TV would not be detected. If the Doppler threshold is set low, micro-Doppler motions, which are away from the macro-Doppler range bins, can also be detected; however, it

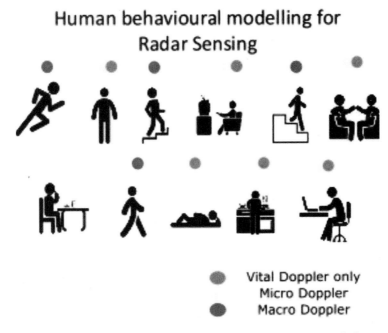

Figure 8.10 Various human activities and corresponding detectable motions [20].

becomes susceptible to false triggers from interference and noise, which means it turns on a light when the space is still unoccupied. Also, since macro-Doppler have large Doppler components while vital-Doppler have a much lower frequency, thus proper sensing of these components under all scenarios require conflicting requirements on PRT and CPI or observation window. Further, macro-Doppler movement at one range bin corrupts the micro-Doppler and vital Doppler signal at neighboring range bins, resulting in false negatives in people counting.

8.4.2 Solution

A signal processing approach to overcome the challenges posed by different properties of the Doppler frequency proposes to process the macro-Doppler, micro-Doppler, and vital-Doppler seperately. Macro-Doppler processing requires closely spaced time samples for estimation, micro-Doppler needs large time spacing and observation windows, and vital-Doppler requires even larger window or frame sizes to estimate minute physiological motions. Furthermore, sensing of macro-Doppler, micro-Doppler, and vital-Doppler motions requires different thresholds, center frequencies, and bandwidths, making separate processing of these signals a desirable approach. The extraction or synthetically creating micro-Doppler frames and vital-Doppler frames from the physical/macro-Doppler frames can be achieved through the frame structure as depicted in Figure 8.11.

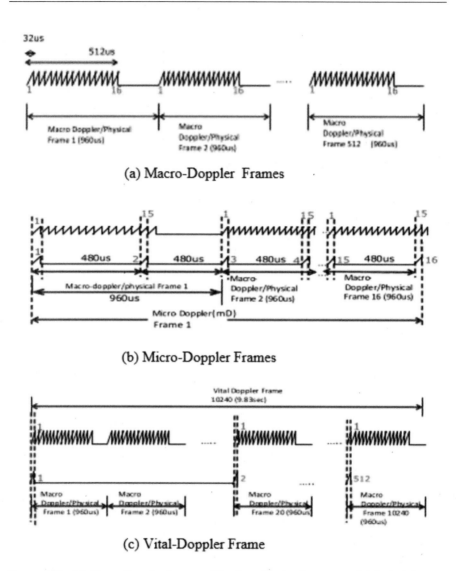

(a) Macro-Doppler Frames

(b) Micro-Doppler Frames

(c) Vital-Doppler Frame

Figure 8.11 (a) Macro-Doppler frames, (b) micro-doppler frames, and (c) vital-doppler frames [20].

In one of the exemplary settings, the physical frame structure with chirp time and pulse repetition interval is set as 32 μs. The physical frame is set to have 16 consecutive chirps, which are collected and processed through a macro-Doppler pipeline. By construction, the maximum Doppler frequency that can thus be detected is 15.625 kHz corresponding to 39.0625 m/s velocity with Doppler resolution of 1.953 kHz. Additionally, as a part of macro-Doppler processing, prefiltering the slow-time data from 2 kHz to 15.6 kHz removes interference from micro-Doppler motions or gestures. Further, subtracting the mean across macro-Doppler slow time removes static targets such as chairs, tables, and inanimate objects. Therefore, this background removal would also eliminate effects from stationary or quasi-stationary humans due to short observation window. The range-macro-Doppler image (RDI) is created by FFT along fast time, followed by slow-time data. The threshold is set adaptively through either CFAR or common adaptive scaled mean. The range bin, which detects macro-Doppler motion, is flagged for person count.

In one of the exemplary settings, the micro-Doppler or arbitrary gestures are detected using a synthetically created micro-Doppler frame, which was created by extracting the first and 15th chirp from the physical/macro-Doppler frame, so that the pulse repetition time becomes 480 μs. The first and 15th chirps are extracted from 8 consecutive frames, making a total of 16 consecutive chirps with a larger PRT, thus lower unambiguous Doppler frequency and better Doppler resolution than that from macro-Doppler frame. By design, therefore, the maximum detectable Doppler is 2.083 kHz corresponding to 2.6 m/s velocity and Doppler resolution of 138.89 Hz. Similar to macro-Doppler processing, the slow-time data is passed through a filter with bandpass frequency from 50 Hz to 2 kHz, followed by mean subtraction for static target removal. The static target removal is followed by RDI generation, over which the adaptive threshold is performed to detect micro-Doppler motions. Once a range bin is detected as exhibiting micro-Doppler or gesture motions it is also flagged for person count.

In the case of vital-Doppler detection, a chirp is extracted every 20th physical frame, thus creating a maximum detectable unambiguous Doppler as 52 kHz. In the vital-Doppler frame, 512 such chirps are collected spanning over 9.83 s observation time. Unlike the macro-Doppler and micro-Doppler processing, the vital detection is done using interferometric techniques. The vital slow-time data is passed through a bandpass filter with frequency over 0.1 Hz to 3 Hz. By interferometric techniques, the phase change over selected range bins are estimated to check if there is a periodic displacement which correspond to that from human breathing or heartbeats [35].

Once the corresponding range bins are flagged as a valid people target from different processing pipelines, the range bins are clustered using the nearest neighbor algorithm or Euclidean clustering based on the maximum width of person to avoid multiple counts from the same target/human. The algorithm flow diagram for separate processing of macro-Doppler, micro-Doppler, and vital-Doppler from their respective frames are presented in Figure 8.12. Furthermore, in a continuous real-time system a count hysteresis is applied to avoid false negatives during transients in between macro/micro-Doppler to vital-Doppler, for example, when a person walking into a room (macro-Doppler) becomes stationary (vital-Doppler). The debouncing logic also helps to counter missed detections or ghost targets intermittently during some frames.

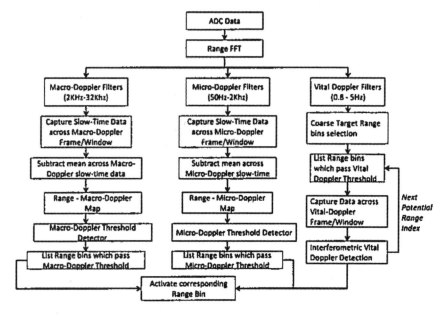

Figure 8.12   People counting algorithm flow [20].

8.4.2.1   Results and Discussions

To evaluate the performance of the proposed signal processing approach to count people in field of view of the radar, some measurements were done under different scenarios in a standard office building, wherein the participating indiviuals exhibited combinations of macro-Doppler, micro-Doppler, and vital Doppler. Since the adaptive threshold for all processing is different, the maximum detectable range of either motions are often different. The measurements included data from meeting rooms, cafeteria, restrooms, and office areas that included desks, chairs, couches, computers, and two interior walls. A total of 60 different cases involving no person, a single person, two persons, and three persons in the field of view were recorded. Table 8.1 summarizes the system parameters that the chip was operated with. The azimuth field of view is around 70° and we operate the chip as a 1 Tx and 1 Rx mode. The total power consumption of the chip is between 10.8 mW to 320 mW, where higher power consumption is for full 2 Tx and 4 Rx mode operation with maximum chirp and bandwidth settings.

Figure 8.13(a) presents the mD range-Doppler image of a human walking into a room and Figure 8.13(b) presents the μD RDI of a human working on a computer. Figure 8.14 depicts the received power profile of a vDoppler component due to breathing/heartbeats from a person sitting idle at 4m to the sensor. The different Doppler components/signatures are evident in these figures.

In the controlled measurement setup, it was ensured that human subjects were within the field of view of the sensor and for the vital-Doppler

Table 8.1
System Parameters Summary [20]

Symbol	Parameter	Value
f_c	Carrier Frequency	60 GHz
NTS	Number of ADC Samples	128
F_s	ADC Sampling frequency	2 MHz
T_c	Chirp Time	32 μs
N_{TX}	Num.Transmit Antennas Used	1
N_{RX}	Num.Recieve Antennas Used	1
B	Total Bandwidth Used	2 GHz
Δ_r	Range resolution	7.5 cm
R_{max}	Maximum unambiguous range	4.8 m
CPI	Coherent Pulse Interval	512 μs
T_{frame}	Frame Time	960 us
Θ_{FOV}	Azimuth Antenna Field of View	70°

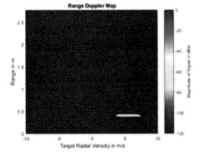 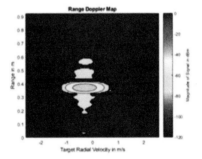

(a) Range Macro-Doppler Map   (b) Range Micro-Doppler Map

Figure 8.13 (a) Range Macro-Doppler Maps of person walking into the room and (b) Range Micro-Doppler Maps of person working on computer [20].

(sitting/standing idle) case the human was facing the sensor by at least 50° to faciliate vital-Doppler sensing by the vital-interferometric pipeline. Table 8.2 presents the confusion matrix of people count for up to three persons using the proposed signal processing solution under 60 controlled scenarios. The proposed solution achieved 100% accuracy in estimating occupancy (i.e., presence or absence of humans). We observed some inaccurate people estimates in cases where three people in a room exhibited macro-Doppler (walking) and macro-Doppler (walking) while another person exhibited only vital-Doppler (sitting idle), respectively. Inaccurate counts were also observed when three people exhibited macro-Doppler (walking), macro-Doppler (walking), and micro-Doppler (working on computer), respectively. Although the parallel processing

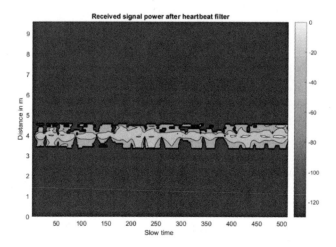

Figure 8.14   Vital-Doppler received power profile [20].

Table 8.2
Confusion Matrix of People Counting Proposed Algorithm
in a Room [20]

		Predicted			
		None	**1 Person**	**2 Persons**	**3 Persons**
	None	15	0	0	0
	1 Person	0	15	0	0
Actual	2 Person	0	0	15	0
	3 Person	0	0	2	13

is designed to separate out the different Doppler components and avoid interference from specifically larger amplitude Doppler components to affect smaller amplitude Doppler components, the signal processing based separation still isn't perfect. Further, the counts of people was incorrect for more than three persons and thus the need for deep learning-based approach.

8.5 People Counting: Deep Learning Approach

Both camera and radar-based people counting systems have their respective benefits and limitations. In this section, we introduce a novel multimodal-cross learning framework to distill high-level feature abstraction learned from camera-based DCNN to radar DCNN so that during testing/inference the radar DCNN alone can deliver a far superior performance compared to its counterpart unimodal radar DCNN in counting people. Unimodal learning refers to the

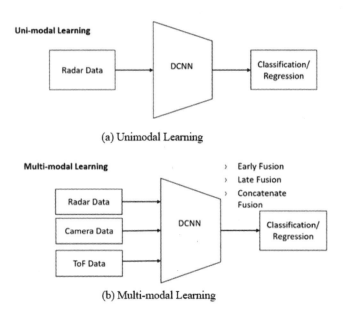

(a) Unimodal Learning

(b) Multi-modal Learning

Figure 8.15 (a) Unimodal learning, wherein only a radar data stream is used for training and inference, and (b) multimodal learning, wherein multidata modality such as radar, cameras, and ToF data streams are used for training and inference.

framework wherein a single modality data stream, such as radar, is used for a particular classification or regression task, as shown in Figure 8.15(a). Multimodal learning refers to the framework wherein multiple sensor data streams, such as camera and radar, are trained together for a common classification or regression task; likewise, in testing/inference the same sensor fusion setup is used as shown in Figure 8.15(b). However, in the case of multimodal cross learning, the radar DCNN and camera DCNN are trained separately or together for a common task and through well-designed loss function, high-level feature abstractions and representations are distilled from the superior DCNN to the inferior DCNN, so that the originally inferior DCNN is capable of delivering as similar performance to the superior DCNN. The superior DCNN could be due to availability of a large training data set, high-resolution data, or much deeper neural network. Figure 8.16 presents the high-level concept of the proposed multimodal cross-learning framework where radar DCNN learns not only from radar data but also supervised representation from camera DCNN trained on a similar task.

Several works have focused on combining information, via multimodal learning, from several modalities either through slow-fusion at lower abstraction

Training

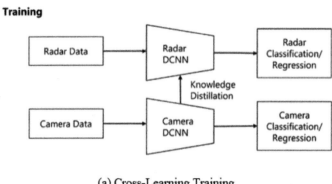

(a) Cross-Learning Training

Testing/Inference

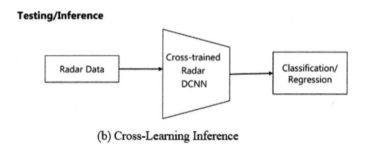

(b) Cross-Learning Inference

Figure 8.16 (a) Training methodology for multimodal cross learning through knowledge distillation from a superior sensor modality's network, and (b) testing or inference setup using cross-learned neural network parameters [41].

layers or late fusion at higher abstraction layers or through concatenation of the extracted features from independent streams [36,37]. Recently, RFPose [38] was proposed wherein the authors propose a camera DCNN to act as a teacher network providing supervision to WiFi-based DCNN to learn artifacts and features that can help WiFi-based DCNN to predict human poses during inference—a task seemingly difficult for WiFi DCNN alone to achieve. In [39], the authors have proposed a correlation-based similarity metric to learn features from one modality to another modality to achieve superior performance for gesture recognition tasks. In this section, we propose a novel multimodal cross-learning framework for people counting application using FMCW radar, which is trained not only from supervised radar data but also learns its parameters through high-level features distilled from camera-based DCNN. We demonstrate the people counting performance of our proposed solution with up to four people counting and detection of more than four people in an indoor environment, which surpasses the performance achievable through its counterpart radar DCNN exploiting supervised radar-only data.

8.5.1 Data Preparation and Processing

8.5.1.1 Range-Angle Image

The intermediate frequency signal from a chirp with NTS $= 128$ number of samples and PN $= 64$ consecutive chirps are collected and arranged in the form of a 2D matrix as PN $\times$ NTS. As a first step, a RDI is generated by subtracting the mean along fast time, followed by 1D FFT along fast time for all the PN chirps to obtain the range transformations. Following which mean across slow time is subtracted followed by 1D FFT to obtain the Doppler transformation for all range bins. The RDI is then processed through an MTI filter to remove reflections from any static targets, such as chairs and furniture in the room. Once the RDI across both received channel $N_{Rx} = 2$ is computed, the RAI is computed through a digital beam-forming algorithm utilizing the derived weights from the angle model as follows:

$$
z_{RAI}(r, \theta) = \sum_{v=-v_{max}}^{v_{max}} \sum_{j=1}^{N_{Rx}} z_{RDI}^{j}(r, v) e^{-j \frac{2\pi d j \sin(\theta)}{\lambda}} \quad \forall -\frac{\theta_{azim}}{2} < \theta < \frac{\theta_{azim}}{2}
$$

(8.10)

where θ is the estimated angle sweeped across the field of view; that is, $-\theta_{azim}/2 < \theta < \theta_{azim}/2$, where θ_{azim} is the half-power beam width, and z_{RDI}^{j} is the complex RDI from the jth receive channel across N_{Rx} receive channels. The first summation in equation (8.10) transforms the RDI across each virtual channel into RDI across the angle space, and the second summation marginalizes across Doppler bins to generate the RAI.

Figure 8.17 presents the preprocessing step, whereby two separate range-angle images are generated using macro-Doppler processing and micro-Doppler processing by using chirps within the physical frame and extracting the first chirp from 32 consecutive physical frames, as explained in detail in Section 8.4. In case of the static person, the human target generates micro-motion dynamics due to breathing or small body movements, resulting in Doppler modulations on the returned signal [40]. Thus, macro-Doppler RAI-channel captures human targets during movements and major motions, while micro-Doppler RAI-channel captures human targets during static or quasi-static scenarios.

Figure 8.18 presents the RAI generated from a macro-Doppler and micro-Doppler processing pipeline. As can be seen, in the case of a static person the RAI generated from the micro-Doppler processing has a clear peak while that from the macro-Doppler processing doesn't, and alternately for moving humans. The two RAIs are fed as separate channels into the deep neural net.

8.5.1.2 Camera Processing (CSR-Net)

In [25], the authors propose a network that can generate highly accurate people counting and provide density heatmaps even in highly congested crowds from a

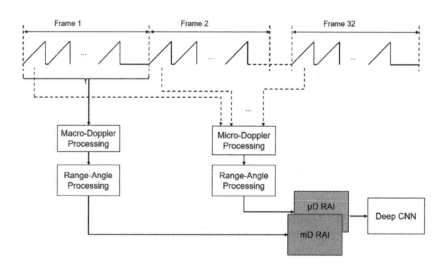

Figure 8.17 Preprocessing pipeline to generate separate range-angle image using micro-Doppler and macro-Doppler components and is fed as separate channels to the deep neural net [41].

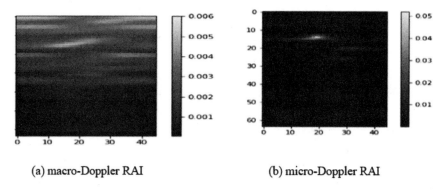

(a) macro-Doppler RAI   (b) micro-Doppler RAI

Figure 8.18 RAI generated from macro-Doppler components and micro-Doppler components for a static person [41].

camera image. The proposed network consists of a modified *VGG-16* network without the classification block, which shrinks down the image by one eighth of its original size. The authors proposed 9 slight modification to the original *VGG-16* to reduce computation requirements, by using only the first 10 layers and having only three pooling layers instead of five. The output of the modified *VGG-16* network is scaled to match the size of input by performing bilinear interpolation with a factor of 8. After the scaling, the output is fed to a dilated convolution network, which is a very critical design aspect as it allows to have a large receptive field using small kernels.

A 2D dilated convolution can be defined as followed:

$$y(m, n) = \sum_{i=1}^{M} \sum_{j=1}^{N} x(m + r \times i, n + r \times j) w(i, j) \qquad (8.11)$$

where $y(m, n)$ is the output of dilated convolution from input $x(m, n)$ and a filter $w(i, j)$ with r being the dilation rate (normal convolution when $r = 1$). The dilated convolution network uses six layers adapting the same 3×3 kernel to a dilated kernel with dilation rate of 2 having 9 progressive number of feature maps as 512, 512, 512, 256, 128, 64. The final camera-based density maps are generated by using a geometry adaptive kernel where each head annotation is blurred with a Gaussian kernel. The kernel is defined as:

$$F(x) = \sum_{i=1}^{N} \delta(x - x_i) \times G_{\sigma_i}(x); \quad \sigma_i = \beta d_i \qquad (8.12)$$

where the $\delta(.)$ is the ground truth and x the target object in it and is convoloved with a Gaussian kernel whose standard deviation is given by multiplication of $\beta = 0.3$ and the average distance d_i of three nearest neighbors.

To adapt the network to work well using our camera setup to model the nonlinear noise, transfer-learning is used by retraining the last two layers of the *CSR-Net* by feeding a few supervised camera images. We refer to this model as adapted *CSR-Net* in the sequel. The adapted CSR-Net generates a density map for a given image input and the count can be done on the output density maps by counting the number of pixels where the value exceeds a defined threshold after clustering. Figure 8.19 depicts the density heatmap output of the adapted *CSR-Net* along with the corresponding camera image for two and three people in the room.

8.5.2 Solution: Framework and Learning

In our proposed solution, a 2D CNN autoencoder is developed wherein the input is the radar RAI and the network is trained to reconstruct the density heatmap, which is generated by the adapted CSRNet from the corresponding synchronized camera image. During training, thus the idea is to learn feature embedding, which captures the representation and semantics derived from both the radar input image as well as the density maps from the camera-based DCNN. The aim of the proposed training framework is to improve the learning process by transferring or distilling the high-level knowledge abstraction (density heatmap) from the camera modality to the radar DCNN, and act as an additional supervision, much like the teacher-student network. To achieve this knowledge distillation from a high accurate network modality, we propose a novel reconstruction loss function, which is a combination of focal-regularized mean square error and cross entropy.

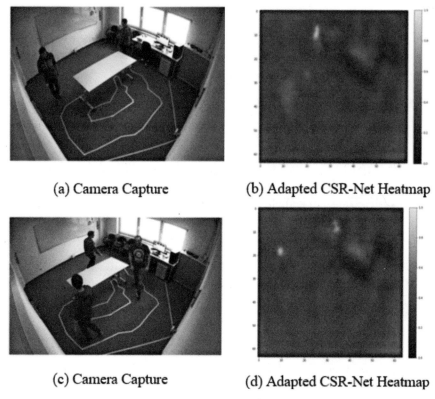

(a) Camera Capture   (b) Adapted CSR-Net Heatmap

(c) Camera Capture   (d) Adapted CSR-Net Heatmap

Figure 8.19   Camera image of (a) two people (c) three people in the field of view, corresponding heatmap from adapted CSR-Net in (b) and (d), respectively [41].

Figure 8.20(a) presents the proposed framework for training the radar RAI to learn feature embedding that is capable of reconstructing a high-quality density heatmap perfectly suitable for the task of people counting. Figure 8.20(b) presents the second step in training, the learned embedding from the autoencoder is fed into a fully connected layer followed by softmax and only the FC layer is retrained through binary cross-entropy loss for the people count class while keeping the encoder weights frozen. In this section, we outline the DCNN autoencoder architecture, the overall loss function, and the final classification step for people counting through classification.

8.5.2.1 Convolutional Autoencoder

For our use case, we use a convolutional autoencoder that would learn to map the radar RAIs generated from the radar data to the density heatmap generated by the adapted *CSRNet* on the corresponding camera images. The encoder part has three consecutive convolutional-pooling layers (conv1-pool1-conv2-pool2-conv3-pool3). Additionally, there are added layers between the first two and last

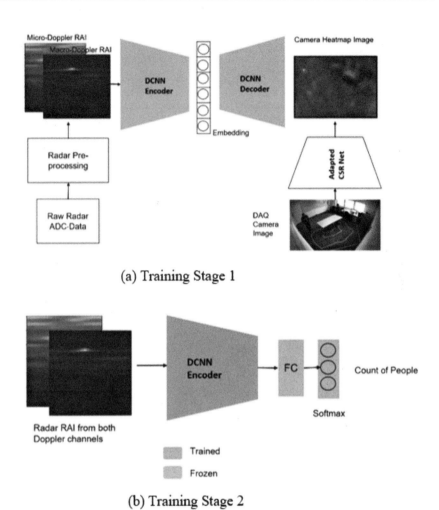

(a) Training Stage 1

(b) Training Stage 2

Figure 8.20 Proposed multimodal cross-learning framework, (a) Stage 1 network is trained to learn camera CSR-Net heatmap, and (b) Stage 2 network with fully connected layer learn to count the number of people [41].

two pooling layers, which act as residual connections for convolutional layers. The decoder part similarly consists of three consecutive convolutional upsampling layers and one additional convolutional layer with sigmoid activation at the end for the reconstruction. Residual add connections are used again within the decoder part, between the upsampling layers. Each convolutional layer output 32 feature maps, with 3 × 3 kernel size and rectified linear unit activations. Pooling layers downsample their input by a factor of two; similarly, the upsampling layers scale up their input by a factor of two. Since the supervised reconstruction image for the autoencoder is high-resolution heatmaps, the network learns feature representation based on high-level feature abstractions (heatmaps) from the camera modality.

8.5.2.2 Loss Function

We propose a loss function based on λ-combination of mean-square error (MSE) and cross entropy (CCE) with data-dependent focal regularization. The MSE error measures how close the reconstructed input X_r is to the original input X in a Euclidean-distance sense. The MSE loss is given as $l_{\text{MSE}}(X, X_r) = 1/N||X - X_r||^2$. The CCE is another suitable loss function that computes how many bits of information are preserved due to reconstruction compared to the ground truth. The $l_{\text{CCE}}(X, X_r) = \sum_k X^k \log X_r^k + (1 - X^k) \log(1 - X_r^k)$.

Now, since in the density heatmap the number of pixels close to zero is much higher compared to the ones which are closer to one, which actually represents the head of the human target, it necessitates the use of some regularization. In the case of MSE, we achieve this by a selective scaling of the cost function based on the threshold of the original pixel in the heatmap, whereas in the case of CCE, we ensure it through a power factor γ, which controls the training focus applied to a particular class of examples. In our case due to the imbalance, a higher focal loss is applied to the pixels crossing a certain threshold adaptively chosen through mean scaling per image.

$$l_{\text{MSE}} = \frac{1}{N_0} w^0 ||X^0 - X_r^0||^2 + \frac{1}{N_1} w^1 ||X^1 - X_r^1||^2 \tag{8.13}$$

$$l_{\text{CCE}} = \sum_{k \notin \Omega} (X^k)^{\gamma_0} \log(X_r^k) + \sum_{k \in \Omega} (1 - X^k)^{\gamma_1} \log(1 - X_r^k) \tag{8.14}$$

$$l_{\text{total}} = l_{\text{CCE}} + \lambda * l_{\text{MSE}} \tag{8.15}$$

where w^0 and w^1 are the selective weights applied to ground-truth pixels, which are less and greater than the adaptive mean, respectively, whereas γ_0 and γ_1 are the power factors that control the selective emphasize applied to ground-truth pixels, which are less and greater than the adaptive mean, respectively. The set Ω represents pixels in X_r that cross a threshold η, which we chose to be 0.4. For our case, $w^1 > w^0$ and $\gamma_1 > \gamma_0$. λ is the weighting parameter applied to control the contribution of the MSE loss to the total loss function; this is a hyperparameter and is computed through cross validation.

8.5.2.3 People Count Classification

Once the autoencoder is trained for the reconstruction task, the feature embedding, which is the output of the encoder unit, is further fed to a fully-connected layer with 10 hidden units with a softmax activation to count people from 1–5 classes, where the fifth class represents more than four people. In this step, the weights of the trained encoder are kept frozen and only the added

FC-layer is trained through binary cross-entropy loss function using the people counts from the training data set as labels in this case. In the first step of training, since the encoder already learns to generate feature embeddings that capture the high-level abstractions and semantics that are capable of classifying people count, a fully connected layer with softmax can readily predict the number/density of people. As an alternative approach, the entire network could be trained end-to-end as a multitask CNN to achieve both the heatmap reconstruction and the people count.

8.5.3 Results and Discussion

The 60-GHz FMCW radar chipset *BGT60TR13C* from *Infineon Technologies AG* was used for evaluation of the proposed deep learning based solution. The frequency chirps with a bandwidth of 1.0 GHz within the 60-GHz band and pulse repetition time of 400 μs were used. The frame time is set to 100 ms, number of ADC samples per chirp NTS = 128, and PN = 64 number of chirps per frame was used.

The chipset *BGT60TR13C* is configured with the system parameters and derived parameters provided in Table 8.3.

The data was collected by syncing four cameras mounted in four corners of a room of size 8m × 10m and four radars placed just below them. The frames per second was set to 10. The recording was performed with the help of 10 individuals who entered or left the room randomly. However, a constraint of a maximum of seven people in the room at any given point of time was imposed. The per frame

Table 8.3
Operating Parameters [41]

Parameters, Symbol	Value
Ramp start frequency, f_{min}	60.5 GHz
Ramp stop frequency , f_{max}	61.5 GHz
Bandwidth, B	1 GHz
Range resolution, δr	15 cm
Number of samples per chirp, NTS	128
Maximum range, R_{max}	9.6 m
Sampling frequency, fs	2 MHz
Chirp time, T_c	64 μs
Chirp repetition time, T_{PRT}	400 μs
Maximum Doppler , v_{max}	3.125 m/s
Number of chirps, PN	64
Doppler resolution, δv	0.0977 m/s
Number of Tx antennas, N_{Tx}	1
Number of Rx antennas, N_{Rx}	2
Elevation θ_{elev} per radar	90°
Azimuth θ_{azim} per radar	130°

people count labeling was done in an automatic fashion by using the adapted CSR-Net. Additionally, a manual scrutiny was performed framewise against the label.

The recording was done in five sets where each set was roughly 5 mins. After each set, different static objects such as tables, chairs, wall mounts, and so on, were added or removed. The total data set includes 59,720 frames with an average of 5,000 frames for each count class. A training and testing split of 80% and 20% on the total data set was performed. Since the aim of the solution is to run on low-commodity hardware, the network architecture for the autoencoder model was found optimal based on its accuracy and model size. During inference, we just need the trained encoder block with the classification block, which takes in input the raw radar data and predicts the count class. The entire inference block has a very small memory footprint of 44 kB, which is an added advantage arising from the proposed multimodal cross learning. The use of only convolution and pooling layers instead of fully connected layers in the encoder block makes the architecture much more robust and fast.

Figure 8.21 presents the camera heatmap generated by the adapted *CSR-Net*, the input radar RAIs, and the reconstructed heatmap at the output of the trained autoencoder for exemplary scenarios of two people and three people. As can be observed from the reconstructed image, the radar autoencoder is capable of transforming the low-resolution radar RAIs into high-resolution heatmaps, which is strikingly close to the ones provided by the adapted *CSR-Net* from the camera input image. Thus, when the embedding representation learned by the trained encoder is used for classification of number of people in its field of view, the deep Net predicts quite accurate people counts.

Figure 8.22(a) presents the confusion matrix of count of people from 1 to 4 and detection of more than four people using unimodal learning, wherein the autoencoder was trained with input from radar RAI and reconstruction to the same radar RAI as well. The compact feature learned by the autoencoder in the next step is used for classifying the number of people. In contrast, Figure 8.22(b) presents the confusion matrix of a count of people using the proposed multimodal cross learning through a camera density heatmap. The same autoencoder architecture is used with different training modalities in both the cases for fair comparison. The unimodal approach is able to reach classification accuracy of 0.86, while the accuracy of the proposed solution is 0.955 on the test data set, demonstrating the superior performance of the proposed solution [41].

8.6 People Detection and Localization: Signal Processing Approach

In human sensing, cognizance or information is to localize or detect where people are in the 2D space. Through a conventional signal processing approach, this can be achieved by generating the range-Doppler image across multiple antennas,

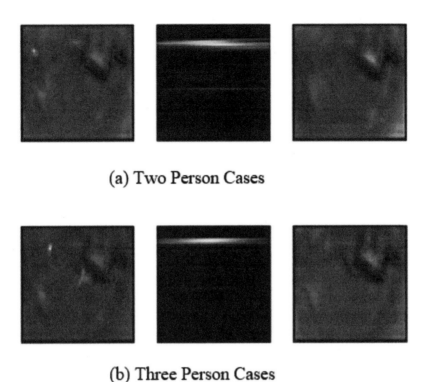

(a) Two Person Cases

(b) Three Person Cases

Figure 8.21 Heatmap from CSR-Net corresponding macro-Doppler RAI and reconstructed heatmap by proposed autoencoder for (a) two people, and (b) three people in the field of view [41].

followed by 2D CFAR-based detection and target clustering. Post target detection in the range-Doppler image, the angle of arrival can be estimated through a monopulse or Capon algorithm, as explained in Chapter 1.

The IF signal from a chirp with NTS number of samples in PN consecutive chirps data are first collected and arranged in the form of a 2D matrix with a dimension of PN × NTS. The RDI is generated in two steps. The first step involves calculating and subtracting the mean along fast time (rows), followed by applying a 1D window function, zero-padding, and 1D FFT along the fast time for all the PN chirps to obtain the range transformations for all chirps in the frame. Then in the second step, the mean along slow time is calculated and subtracted, followed by a 1D window function, zero-padding, and 1D FFT along slow time (columns) to obtain the Doppler transformation for all range bins. Once the RDI across multiple antennas in a uniform linear array are generated, the following algorithms are applied to generate processed RDI with target detections -

1. *Maximal ratio combining (optional):* Following the RDI creation across all virtual recieve channels, MRC is optionally applied to combine the RDIs

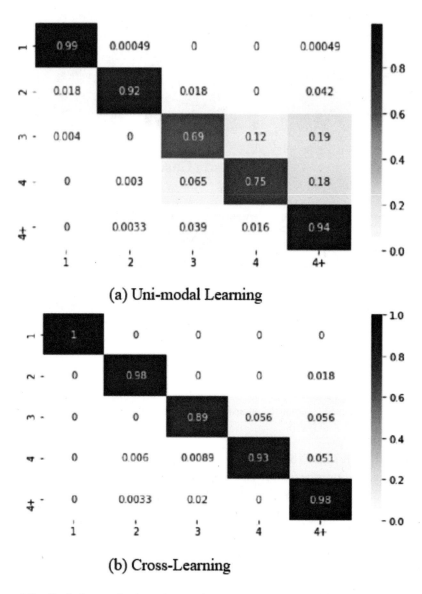

(a) Uni-modal Learning

(b) Cross-Learning

Figure 8.22 Confusion matrix of people counting using radar data with (a) deep convolutional neural network, whose parameters are learned through radar-only data, and (b) deep convolutional neural network whose parameters are learned through camera-based cross learning with radar data [41].

from multiple antennas through a weighted mean to create a single high-fidelity RDI. Combining the multiple RDIs into a single RDI through MRC helps in increasing the signal-to-noise ratio and improve diversity (i.e., occlusion or missed target detection in one channel is offset by detection in another channel). One of the approaches to compute the weights for combining the RDIs are through their respective estimated SNR. The effective RDI can be computed as

$$\text{RDI}_{mrc} = \frac{\sum_{rx=1}^{N_{Rx}} g^{rx} |\text{RDI}^{rx}|}{\sum_{rx=1}^{N_{Rx}} g^{rx}} \tag{8.16}$$

where RDI^{rx} is the complex RDI of the rxth receive channel, and the gain is adaptively calculated as

$$g^{rx} = (\text{NTS.PN}) \frac{\max\{|\text{RDI}^{rx}|^2\}}{\left(\sum_{l=1}^{NTS} \sum_{m=1}^{PN} |\text{RDI}^{rx}(m,l)|^2 - \max\{|\text{RDI}^{rx}|^2\}\right)} \tag{8.17}$$

where $\max\{.\}$ represents the maximum value from the 2D function, and g^{rx} represents the estimated SNR at rth receive channel. Thus, the $(\text{PN} \times \text{NTS} \times N_{Rx})$ RDI tensor is transformed into $(\text{PN} \times \text{NTS})$ RDI matrix. We refer to this effective RDI as raw RDI in the sequel; alternately, just one RDI from channel can be used.

2. *Adaptive detection algorithm:* The raw RDI is fed into a CFAR detection algorithm to generate a mask of 0s and 1s, where 1s indicate presence of a detected target. The target detection problem on the RDI can be expressed as

$$\text{RDI}_{\text{det}}(cut) = \begin{cases} 1 & \text{if } |\text{RDI}_{\text{mrc}}(cut)|^2 > \eta\sigma^2 \\ 0 & \text{if } |\text{RDI}_{\text{mrc}}(cut)|^2 < \eta\sigma^2 \end{cases} \tag{8.18}$$

The CFAR detector is based on Neyman Pearson detection, where the optimization problem is to maximize the probability of detection subject to probability of false alarm being less than a predefined value, such as 0.001. By assuming Gaussian noise or interference, the scaling factor η in CA-CFAR is provided as

$$\mu = N(P_{fa}^{-1/N} - 1) \tag{8.19}$$

where P_{fa} is the probability of false alarm and N is the window size or reference cells used for estimating the noise power σ^2. CA-CFAR is the most common detector used in the case of a point target under varying Gaussian noise/interference. In the case of humans, the target reflections are typically spread along range and Doppler domain this leads to poor performance (missed detections) since the targets also appear in the reference cells making the estimated noise power inaccurate. Additionally, CA-CFAR elevates the noise threshold near a strong target, thus occluding or masking nearby weaker targets. Thus, for detecting doubly spread targets such as humans, OS-CFAR is a better adaptive detector. OS-CFAR not only addresses such outliers but also alleviates the need of guard cells. In the case of OS-CFAR, instead of the mean power in the reference cells, the kth ordered data is selected as the estimated noise variance, σ^2, and this works better for outliers since its a known fact that median is robust to outliers compared to mean (used in CA-CFAR).

3. *DBSCAN:* Contrary to a point target, in the case of doubly extended targets, the output of the detection algorithm is not a single detection in the RDI for a target but a set of detected points across the range and Doppler. Thus, a clustering algorithm is required to group the detections from a single target, based on its size, as a single cluster. For this, the DBSCAN algorithm is used, which is the most common and effective unsupervised learning algorithm. Given a set of target detections from same and multiple targets in the RDI, DBSCAN groups detections that are closely packed together, while at the same time removing as outliers detections that lie alone in low-density regions.

   To do this, DBSCAN classifies each point as either core point, edge points, or noise. Two input parameters are needed for the DBSCAN clustering algorithm, the neighborhood radius d, and the minimum number of neighbors, M. A point is defined as core point if it has at least $M - 1$ neighbors (i.e., points within the distance d). An edge point has less than $M - 1$ neighbors, but at least one of its neighbors is a core point. All points that have less than $M - 1$ neighbors and no core point as a neighbor do not belong to any cluster and will be classified as noise [42]. Based on the chosen set of parameters, this often results in true target splits or causes two neighboring targets to merge.

Once the targets are detected, using the range-Doppler index across multiple antennas, we can estimate the target's angle, which can be used to translate target's polar coordinates to Cartesian coordinates in 2D space.

8.7 People Detection and Localization: Deep Learning Approach

8.7.1 Challenges

Detection of humans in indoor environments pose multiple challenges such as multipath reflections, ghost targets, target occlusion, merging targets, and split targets. Multipath reflections and ghost targets occur due to reflections from static objects like walls, chairs, or tables to the human target and back to the radar sensor, or conversely, reflection from human to these static targets and back to the radar sensor. Additionally, ghost targets also appear due to other moving or vibrating targets in the room, such as curtains or windowpanes. Such multipath reflections and ghost targets appear at the output of 2D-CFAR algorithm as

 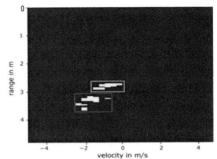

(a) Indoor Room Environment

(b) Processed RDI with a true target (high intensity)
and ghost target (low intensity)

Figure 8.23   (a) Indoor room environment with a human walking around in the room, and (b) corresponding processed RDI as sensed by the radar through signal processing algorithm pipeline MRC, OS-CFAR, and DBSCAN. Processed RDI depicts the true target (in high intensity window) and ghost targets (in low intensity window) due to reflections from walls and other objects [32].

false positives and are erroneously treated as a valid human target. Figure 8.23(a) presents an indoor environment setup wherein a human walks close to the wall, Figure 8.23(b) presents the target detected or processed RDI using conventional signal processing pipeline explained in an Section 8.6, where the true target is marked in high intensity window (closer in range) and ghost targets and multipath artifacts are marked in low intensity window (farther in range).

Another source of error in target detection is due to occlusion of one human by another human in the line of sight to the radar sensors. Furthermore, occlusions can also occur if the reflections from the target are weakened through other RF phenomenon, such as translation range migration, rotational range migration, or speckle [44]. Conventional algorithms through 2D OS-CFAR fail to detect the true target in such scenarios, leading to false negatives.

The other source of error due to merging targets and split targets are due to nonoptimal parameters for clustering algorithms. With DBSCAN clustering specifically, clusters may merge if the neighborhood radius is set too high. However, when set too low, arms, legs, or the head of a human target may be recognized as separate targets. Based on the indoor environment and activity of the human, the radar response from target will have varying points on the range-Doppler domain. Thus, a fixed set of DBSCAN parameters (neighborhood radius) that would work optimally for all practical scenarios is not possible, and adaptively controlling the parameters cannot be achieved easily.

In the efforts to overcome such bottlenecks and challenges posed by traditional processing pipeline with OS-CFAR and DBSCAN, a deep residual U-Net model is proposed to process the raw RDIs into target detected processed RDI. The comparison of the state-of-art signal processing approach with the

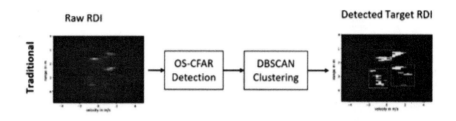

(a) Traditional Processing Pipeline

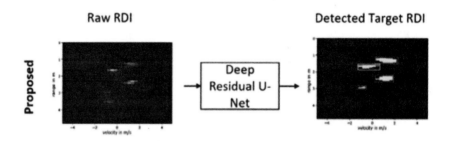

(b) Proposed Processing Pipeline

Figure 8.24   (a) Traditional processing pipeline with OS-CFAR and DBSCAN to generate target detected RDI, and (b) processing pipeline using proposed deep residual U-Net to suppress ghost targets, multipath reflections, mitigate target occlusions, and achieve accurate clustering [32].

proposed deep residual U-Net is presented in Figure 8.24(a) and Figure 8.24(b), respectively. Deep U-Net architectures have been proposed in computer vision community for image segmentation problems; however, in this context deep U-Net model architectures have been repurposed to deal with the problem of missed detection, ghost targets, incorrect target splits, and target mergers, and is described in detail in the following sections.

8.7.2   Architecture and Learning

The proposed deep residual U-Net model contains an encoder and decoder part, each with three depths as illustrated in Figure 8.25. Each block in the encoder path contains two 3×3 convolutions followed by a nonlinear ReLu activation function and a 3×3-max pooling with strides of two along range-Doppler dimension. Correspondingly, each block in the decoder path contains upconvolution or transposed convolution of 2×2 with strides of two in range-Doppler dimension followed by ReLu activation function and two 3×3 convolutions, each followed by a ReLu. The transpose convolutions (or upconvolutions) are implemented

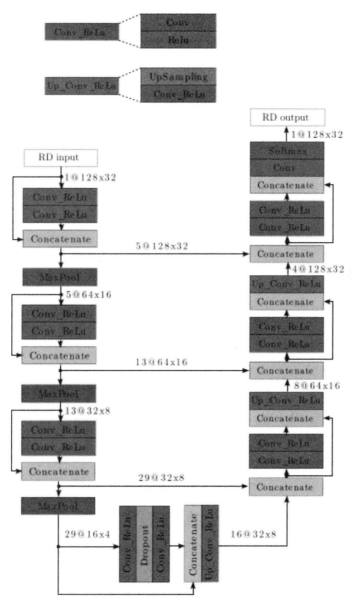

Figure 8.25 Proposed architecture for a three-depth network. Each box corresponds to one or more layers. The number of feature channels with their x and y size is provided throughout the network [32].

as an upsampling layer followed by a convolutional layer. To improve learning, shortcut connections are added from layers of equal resolution in the encoder path to provide the high-resolution features to the decoder path as in the standard U-Net architecture. The bridge connection between the encoder and the decoder network consists of two convolutional layers, each followed by a ReLu with dropout of 0.5 set during training. As well, residual connections in both encoder and decoder paths are applied to improve learning from one layer to a deeper layer. In the last layer, a 1×1 convolution reduces the number of output channels to the number of classes which is 2, target present/absent, in our case. The architecture has 40,752 trainable parameters. The input to the deep residual U-Net is a $128 \times 32 \times 1$ raw RDI, and the output is the processed RDI of the same dimension, $128 \times 32 \times 1$, with pixel values between 0 and 1, where 1 denotes higher probability of detection/presence of target in that pixel.

8.7.2.1 Data Set

The ground-truth data set for training the neural network model is generated by processing the raw RDI through a traditional signal processing algorithm pipeline of OS-CFAR and DBSCAN. To achieve a reliable ground-truth data set, the parameters for detection and clustering algorithms was set to achieve probability of detection close to 100%, compromising on probability of false alarm and resulting in many ghost targets and multipath reflections. This means decreasing the CFAR scaling factor in the case of target occlusion, and reducing/increasing the maximum neighbor distance for cluster detection with DBSCAN in case of merged/separated targets. During the measurements, a synchronized camera was used to record the corresponding scene. The camera data was used for manual annotation of target detections in case of any remaining target occlusion due to other humans or static humans close to a wall, and manual elimination of target detections in case of ghost targets and multipath reflections, thus amounting to application of an artificial correction filter. The described data preparation process is relatively simple for one-target measurements, as the correct cluster is generally the one closest in range to the radar sensor and gets complicated with more people. The (raw RDI, processed RDI) data set comprises from one up to four humans from multiple rooms, with different configurations of static targets, walls, and furniture.

Since generating a labeled data set for more than one person is a tedious task, data augmentation was used to enhance the data set during training and achieve generalization of the model. Data sets for one and two human targets were synthetically superimposed often through translation and rotation to create data sets for three and four human targets, where generating high accurate ground truth data is difficult. Some caution needs to be exercised, otherwise the network can overfit on the number of possible target positions.

8.7.2.2 Loss Function and Parameters

Given a set of training raw RDIs and the corresponding ground truth processed RDIs I_i, P_i, the training procedure estimates the parameters θ of the network, $h(\theta)$, such that the model generalizes to detect accurate processed RDIs. For training the model, a weighted combination of focal loss [43] and hinge loss is used as given by the following set of equations:

$$HL(p) = 1 - y(2p - 1)$$
$$FL(p_t) = \left(1 - p_t\right)^\gamma \log(p_t)$$
$$L(p_t) = \alpha \left[FL(p_t) + \eta HL(p)\right] \tag{8.20}$$

The variables $y \in \{\pm 1\}$, $p \in [0, 1]$, and $p_t \in [0, 1]$, specify the class label, the estimated probability for the class with label $y = 1$, and the probability that a pixel was correctly classified, as defined by

$$p_t = \begin{cases} p & \text{if } y = 1 \\ 1 - p & \text{otherwise} \end{cases} \tag{8.21}$$

The parameters γ, η, and α influence the shape of the focal loss, the weight of the hinge loss, and the class weighting, respectively. The hyperparameters $\gamma = 2$, $\eta = 0.15$, and $\alpha = 0.25$ were found to result in an optimal trained model. The rationale to use focal loss is to handle class imbalance adaptively as the training progresses. In the processed RDIs, the number of pixels with 0 values are high compared to the number of pixels with a value of 1. The focal loss with parameter γ focuses or adds more cross-entropy loss on the hard examples (i.e., the ones that have been incorrectly estimated), while defocusing cross entropy loss on the easy examples (i.e., the ones that have been correctly estimated), thus enabling the network to correctly classify even the rare class (i.e., the detection class 1). The conventional cross-entropy loss would otherwise learn the degenerate model, where it learns all 0s and still maintains a good loss. Further, the hinge loss, being a combination of $L2$-norm and $L1$-norm, adds a mean square error and mean absolute error between the true processed RDI and predicted processed RDI. The focal loss is added as a loss function through factor η, which was chosen in such a way that the focal loss dominates the training for the first few epochs before the hinge loss becomes relevant. The neural network is initialized by drawing from a normal distribution with zero mean and standard deviation of 10^{-2}. The respective biases were initialized with samples drawn from a normal distribution also but with a mean of 0.5 instead of zero. The Adam optimization was used with learning rate of 0.001, and epsilon of 10^{-7} for training.

8.7.3 Results and Discussion

The most common metric to evaluate the detection performance is to evaluate the radar ROC, which plots the probability of detection with respect to probability of false alarm, and the quantitative performance metric is to compute the corresponding area under curve (AUC). In this case of target detection, classification accuracy can be misleading since the absence class (0) is much higher than the presence class (1), and thus the F1 score that combines precision and recall is a better comparison metric. Precision helps when the costs of false positives (i.e., ghost targets or multipath reflections), are high. Recall helps when the cost of false negatives (i.e., target occlusions) are very high. A good F1 score, close to 1, means low false positives (i.e., ghost target detection), and low false negatives (i.e., target occlusion), thus indicating correct target detections with minimal false triggers.

Infineon's FMCW radar chipset *BGT60TR24* was used for evaluation and collecting data sets, and chipset was configured with system parameters provided in Table 8.4. Each chirp has NTS = 256 samples, which represents the DAC/ADC samples, and a bandwidth of $B = 4$ GHz with chirp duration of $T_c = 261$ μs was used. PN = 32 number of chirps per frame were used with pulse repetition time of $T_{PRT} = 520$ μs. Owing to the chosen setting, the theoretical range resolution $\delta r = \frac{c}{2B} = 3.75$ cm and the maximum theoretical range is $R_{max} = (NTS/2) \times \delta r = 4.8$ m is available. The theoretical maximum unambiguous velocity $v_{max} = 4.8$ m/s and the minimum resolvable velocity is $\delta v = 0.3$ m/s. One transmit antenna and three receive antennas were enabled for data collection.

The test data contains a total of 2,000 raw RDIs to evaluate the performance of the neural network in comparison to a traditional signal processing approach. The test set contains up to four human targets from different room configurations performing regular human activities. It is worth noting here that if the MTI step to cancel static targets in the field of view is also attempted to be processed

Table 8.4
Operating System Parameters [32]

Parameters, Symbol	Value
Ramp start frequency, f_{min}	58 GHz
Ramp stop frequency , f_{max}	62 GHz
Bandwidth, B	4 GHz
Range resolution, δr	3.75 cm
Number of samples per chirp, NTS	256
Maximum range, R_{max}	4.8 m
Sampling frequency, fs	2 MHz
Chirp time, T_c	261 μs
Chirp repetition time, T_{PRT}	520 μs
Maximum Doppler , v_{max}	4.8 m/s
Number of chirps, PN	32
Doppler resolution, δv	0.3 m/s
Number of Tx antennas, N_{Tx}	1
Number of Rx antennas, N_{Rx}	3
Elevation θ_{elev} per radar	70°
Azimuth θ_{azim} per radar	70°

through the neural network model, this leads to overfitting and the model doesn't generalize. Its most likely due to infinite possibilities of configuring the static targets in the field of view with varying distances, spread, and intensities. Thus, the preferred approach is to remove the static targets through MTI before feeding into the neural network.

For evaluation of the probability of detection and probability of false alarm was performed by comparing the processed RDI through signal processing approach and deep learning approach with the ground-truth processed RDI. Arising from radar sensor behavior and accounting for manual errors in preparing the ground-truth processed RDI, it is likely that only parts of each human are classified as "target present" in the RDI. To account for such variations, when defining detections and false alarms, small positional errors and variations in the cluster size are accounted for. This is done by computing the center of mass for each cluster in the labeled processed RDIs and the predicted processed RDIs, so that the distance between the cluster centers of the mass of the predicted processed RDI and the labeled processed RDIs is less than 20 cm in range, and 1.5 m/s in velocity.

Table 8.5 presents the performance of the proposed approach in terms of F1 score in comparison to the signal processing approach. Results for the proposed approach are shown for a depth-three network (NN_d3) as shown in Figure 8.25 and also for a deeper network with five residual blocks (NN_d5) in both encoder and decoder paths are presented. The proposed network (NN_d3) achieved an F1 score of 0.89, while the conventional signal processing approach only achieved a 0.71 F1-score. A deeper U-Net with five layers (NN_d5) achieved a slightly better F1 score of 0.91 at the expense of larger model size.

The ROC curve of the proposed residual U-Net (NN_d3) and a deeper network (NN_d5) in comparison to conventional signal processing is presented in Figure 8.26. The ROC curves for the depth-three and the depth-five networks are achieved through varying the threshold that is applied on the output probabilities of the predicted processed RDIs. For the signal processing approach, the scaling factor η is varied to obtain the ROC curve for signal processing approach. The curves are extrapolated for detection probabilities close to one. As depicted from the ROC curve, the proposed residual U-Net architecture provides a much better

Table 8.5

Comparison of Detection Performance of Traditional Pipeline, Proposed U-Net with Depth of three and Depth of five Architectures for People Detection in RDI [32]

Approach	Description	F1-Score	Model Size
Traditional	OS-CFAR with DBSCAN	0.71	-
Deep Residual U-Net (depth 3)	Proposed loss	0.89	616 kB
Deep Residual U-Net (depth 5)	Proposed loss	0.91	2.8 MB

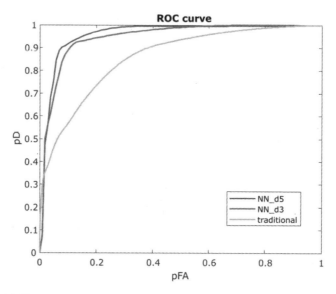

Figure 8.26   ROC comparison between the proposed residual deep U-Net with a traditional signal processing approach [32].

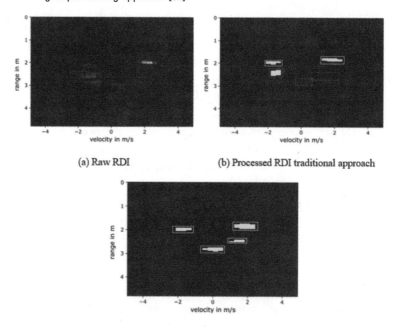

(a) Raw RDI   (b) Processed RDI traditional approach

(c) Processed RDI proposed approach

Figure 8.27   (a) Raw RDI image with three human targets, (b) processed RDI using traditional approach wherein one split target and two targets are occluded, and (c) processed RDI using the proposed approach wherein all targets are detected accurately without false alarms or occlusions [32].

AUC performance compared to the conventional signal processing pipeline (i.e., a higher detection probability for a given probability of false alarm).

Figures 8.27(a), 8.27(b), and 8.27(c) present the raw RDI, processed RDI through the signal processing approach and processed RDI through the proposed residual U-Net, respectively, for a target scenario. The predicted processed RDI values lie between 0 and 1, and for the plot a hard threshold of 0.5 was used for detection of the target class. With the signal processing approach one target at around 2m distance and -2 m/s in velocity was incorrectly split into two separate targets owing to incorrect DBSCAN parameters. As well, two targets between 2 and 3 meters were missed from detection. The predicted processed RDI was able to detect all four targets correctly without target splits or occlusion (miss).

Figures 8.28(a), 8.28(b), and 8.28(c) present the raw RDI, processed RDI through the signal processing approach and processed RDI through the proposed residual U-Net, respectively, for another target scenario. In this case, the problem arising occlusion the conventional signal processing approach missed detecting

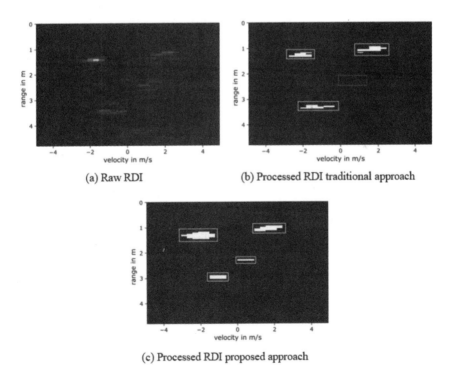

(a) Raw RDI   (b) Processed RDI traditional approach

(c) Processed RDI proposed approach

Figure 8.28 (a) Raw RDI image with four human targets, (b) processed RDI using the traditional approach wherein one target is occluded, and (c) processed RDI using the proposed approach wherein all targets are detected accurately without false alarms or occlusions [32].

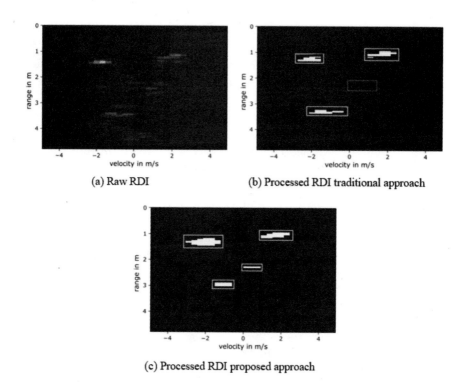

(a) Raw RDI   (b) Processed RDI traditional approach

(c) Processed RDI proposed approach

Figure 8.29   (a) Raw RDI image with four human targets, (b) processed RDI using the traditional approach wherein two targets are merged into one by the DBSCAN clustering algorithm, and (c) processed RDI using the proposed approach wherein all targets are detected accurately and with proper clustering [32].

one target at around a 3-meter distance, while the neural network approach was able to reliably detect all the targets.

Further, in Figure 8.29(b), two detected targets around the 1-meter mark are too close together for the DBSCAN clustering algorithm and thus are grouped as a single target although they are from two different targets. In contrast, in Figure 8.30(b) one target at around 4 meters to the radar sensor was wrongly detected due to target split, again due to the DBSCAN clustering parameter. Nevertheless in both cases, the U-Net based approach correctly displays all the distinct targets as presented in Figures 8.29(c) and 8.30(c).

From our experiments, we have observed that the proposed approach excels in discarding multipath reflections and ghost targets caused by the reflections from static objects in the scene, does well in preventing splitting or merging targets, but does not really show much improvements for the case of occluded targets if many humans are in front of each other. This is most likely due to the fact that the occlusion results in negligible signal in the raw RDI itself, thus

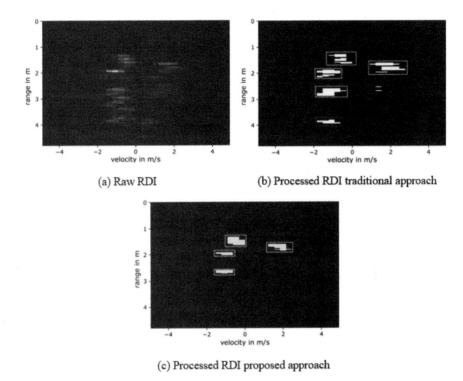

(a) Raw RDI   (b) Processed RDI traditional approach

(c) Processed RDI proposed approach

Figure 8.30   (a) Raw RDI image with four human targets, (b) processed RDI using the traditional approach wherein a ghost target appears, and (c) processed RDI using the proposed approach wherein all targets are detected without false alarms or occlusion and accurate clustering [32].

the neural network fails to recover occluded targets correctly in the predicted processed RDIs.

8.8 Future Work and Direction

In the case of people counting, approaches combining meta-learning and cross learning can be used to improve the count classification further. Additionally, more sophisticated camera heatmap models than CSRNet with an improved cross-learning pipeline in the form of a teacher-student network can help increase the classification count and help model generalization.

In the context of human localization, the U-Net model being sensitive to sensor noise and variation can be replaced by a variational autoencoder to improve the robustness of the overall system. Further, an end-to-end network to input raw ADC data and including parameter estimation directly as an output is another interesting research direction.

8.9 Problems

1. What are the specific benefits of using people sensing and counting applications using radars over a vision-based system?
2. How does MTI filtering in the signal processing pipeline help in terms of deep learning?
3. What kind of weight initialization procedure for sinc filters can be followed to achieve faster convergence?
4. What would be the influence of focal loss over a well-class-balanced data set?
5. Describe a deep learning architecture that can output both the amplitude of RDI and the phase of RDI. State where such a neural network can be used in radar processing.
6. In a meeting room scenario, which radar data representation would be the most important and why?
7. Discuss other applications that can be built using the cross-learning approach. Discuss the training approach, deep learning architectures, and loss function to be used.
8. What is the importance of having residual connections in a U-Net?
9. Compare hinge loss and MSE loss.

References

[1]   *Annual Energy Outlook 2011 with Projections to 2035*, U.S. Energy Information Administration, Independent Statistics and Analysis, http://www.eia.gov/forecasts/aeo/pdf/0383er(2011).pdf.

[2]   Yavari, E., C. Song, V. Lubecke, and O. Boric-Lubecke, "System-on-Chip Based Doppler Radar Occupancy Sensor," in *Annual International Conference of the Engineering in Medicine and Biology Society, EMBC*, August 2011, pp. 1913–1916.

[3]   EPRI, "Occupancy Sensors: Positive On/Off Lighting Control," EPRIBR-100323, Electric Power Research Institute, Palo Alto, CA, 1994.

[4]   G. Vishal and N. K. Bansal, "Smart Occupancy Sensors to Reduce Energy Consumption," *Energy Buildings*, Vol. 32, No. 1, 2000, pp. 81–87.

[5]   Yavari, E., A. Lee, K. Pang, N. A. McCabe, and O.Boric-Lubecke, "Radar and Conventional Occupancy Sensors Performance Comparison," in *Asia-Pacific Microwave Conference (APMC)*, IEEE, 2014, pp. 444–446.

[6]   Yavari, E., C. Song, V. Lubecke, and O. Boric-Lubecke, "Is There Anybody in There? Intelligent Radar Occupancy Sensors," *IEEE Microwave Magazine*, Vol. 15, No. 2, 2014, 2014, pp. 57–64.

[7]   Butler, W., P. Poitevin, and J. Bjomholt, "Benefits of Wide Area Intrusion Detection Systems Using FMCW Radar," in *Proc. 41st Annu. IEEE Int. Carnahan Conf. Secur. Technol.*, 2007, pp. 176–182.

[8]   Lien, J., et al., "Soli: Ubiquitous Gesture Sensing with Millimeter Wave Radar," *ACM Transactions on Graphics (TOG)*, Vol. 35, No. 4, July 2016.

[9]   Hazra, S., and A. Santra, "Robust Gesture Recognition Using Millimetric-Wave Radar System," *IEEE Sensors Letters*, Vol. 2, No. 4, December 2018.

[10] Arsalan, M., and A. Santra, "Character Recognition in Air-Writing Based on Network of Radars for Human-Machine Interface," *IEEE Sensors Journal*, Vol. 19, No. 19, 2019, pp. 8855–8864.

[11] Will, C., et al., "Human Target Detection, Tracking, And Classification Using 24 GHz FMCW Radar," *IEEE Sensors Journal*, Vol. 19, No. 17, 2019, pp. 7283–7299.

[12] Gurbuz, Sevgi Zubeyde, and Moeness G. Amin. "Radar-based human-motion recognition with deep learning: Promising applications for indoor monitoring." IEEE Signal Processing Magazine 36.4 (2019): 16-28.

[13] Munoz-Ferreras, J. M., Z. Peng, R. Gomez-Garcia, and C. Li, "Review on Advanced Short-Range Multimode Continuous-Wave Radar Architectures for Healthcare Applications," *IEEE Journal of Electromagnetics, RF and Microwaves in Medicine and Biology*, Vol. 1, No. 1, 2017, pp. 14–25.

[14] Adib, F., H. Mao, Z. Kabelac, D. Katabi, and R. C. Miller, "Smart Homes that Monitor Breathing and Heart Rate,"in *Proceedings of the 33rd Annual ACM Conference on Human Factors in Computing Systems*, April 2015, pp. 837–846.

[15] Aardal, O., Y. Paichard, S. Brovoll, T. Berger, T. S. Lande, and S. E. Hamran, "Physical Working Principles of Medical Radar," *IEEE Transactions on Biomedical Engineering*, Vol. 60, No. 4, 2013, pp. 1142–1149.

[16] Kim, Y., S. Ha, and J. Kwon, "Human Detection Using Doppler Radar Based on Physical Characteristics of Targets," *IEEE Geoscience and Remote Sensing Letters*, Vol. 12, No. 2, 2015, pp. 289–293.

[17] Li, C., V. M. Lubecke, O. Boric-Lubecke, and J. Lin, "A Review on Recent Advances in Doppler Radar Sensors for Noncontact Healthcare Monitoring," *IEEE Transactions on Microwave Theory and Techniques*, Vol. 61, No. 5, 2013, pp. 2046–2060.

[18] Gennarelli, G., G. Ludeno, and F. Soldovieri, "Real-Time Through-Wall Situation Awareness Using a Microwave Doppler Radar Sensor," *Remote Sensing*, Vol. 8, No. 8, 2016, p. 621.

[19] Wang, Y., Q. Liu, and A. E. Fathy, "CW and Pulse-Doppler Radar Processing Based on FPGA for Human Sensing Applications," *IEEE Transactions on Geoscience and Remote Sensing*, Vol. 51, No. 5, 2013, pp. 3097–3107.

[20] Santra, A., R. V. Ulaganathan, and T. Finke, "Short-Range Millimetric Wave Radar System for Occupancy Sensing Application," *IEEE Sensors Letters*, Vol. 2, No. 3, 2018, p. 14.

[21] Wang, C., H. Zhang, L. Yang, S. Liu, and X. Cao, "Deep People Counting in Extremely Dense Crowds," in *Proceedings of the 23rd ACM International Conference on Multimedia. ACM*, 2015, pp. 1299–1302.

[22] Boominathan, L., S. S. Kruthiventi, and R. V. Babu, "Crowdnet: A Deep Convolutional Network for Dense Crowd Counting," in *Proceedings of the 24th ACM International Conference on Multimedia. ACM*, 2016, pp. 640–644.

[23]   Shi, Z., L. Zhang, Y. Liu, et al. "Crowd Counting withDeep Negative Correlation Learning," in *Proceedings of the IEEE Conference on Computer Vision and Pattern Recognition*, 2018, pp. 5382–5390.

[24]   Zou, Z., X. Su, X. Qu, and P. Zhou, "DA-Net: Learning the Finegrained Density Distribution with Deformation Aggregation Network," *IEEE Access*, Vol. 6, 2018, pp. 60745–60756.

[25]   Li, Y., X. Zhang, and D. Chen, "CSRNet: Dilated Convolutional Neural Networks for Understanding the Highly Congested Scenes," in *Proceedings of the IEEE Conference on Computer Vision and Pattern Recognition*, 2018, pp. 1091–1100.

[26]   Gao, C., P. Li, Y. Zhang, J. Liu, and L. Wang, "People Counting Based on Head Detection Combining Adaboostand CNNin Crowded Surveillance Environment," *Neurocomputing*, Vol. 208, 2016, pp. 108–116.

[27]   Choi, Jeong Woo, Sung Sik Nam, and Sung Ho Cho. "Multi-human detection algorithm based on an impulse radio ultra-wideband radar system." IEEE Access 4 (2016): 10300-10309.

[28]   Choi, J. W., D. H. Yim, and S. H. Cho, "People Counting Based on an IR-UWB Radar Sensor," *IEEE Sensors Journal*, Vol. 17, No. 17, 2017, pp. 5717–5727.

[29]   He, J., and A. Arora, "A Regression-Based Radar-Mote System for People Counting," in *2014 IEEE International Conference on Pervasive Computing and Communications (PerCom)*, 2014, pp. 95–102.

[30]   Yang, X., W. Yin, and L. Zhang, "People Counting Based on CNN Using IR-UWB Radar," in *2017 IEEE/CIC International Conference on Communications in China (ICCC)*, 2017, pp. 1–5.

[31]   Yang, X., W. Yin, L. Li, and L. Zhang, "Dense People Counting Using IR-UWB Radar with a Hybrid Feature Extraction Method," *IEEE Geoscience and Remote Sensing Letters*, Vol. 16, No. 1, 2018, pp. 30–34.

[32]   Stephan, M., and A. Santra, "Radar-Based Human Target Detection Using Deep Residual U-Net for Smart Home Applications," in *Proceedings of the 18th IEEE International Conference on Machine Learning Applications (ICMLA)*, 2019.

[33]   Santra, A., I. Nasr, and J. Kim, "Reinventing Radar: The Power of 4D Sensing," *Microwave Journal*, Vol. 61, No. 12, 2018, pp. 26–38.

[34]   Van Trees, H. L., *Detection, Estimation, and Modulation Theory*, Part I, John Wiley & Sons, 2004.

[35]   Avik, S., V. U. Raghavendran, F. Thomas, B. Ashutosh, W. Jungmaier Reinhard, T. Saverio, and N. Dennis, "Short-Range Multi-Mode Continuous-Wave Radar for Vital Sign Measurement and Imaging," in *IEEE Radar Conference (RadarConf)*, 2018.

[36]   Wu, J., J. Cheng, C. Zhao, and H. Lu, Fusing Multi-Modal Features for Gesture Recognition," in *Proceedings of the 15th ACM on International Conference on Multimodal Interaction. ACM*, 2013, pp. 453–460.

[37]   Neverova, N., C. Wolf, G. Taylor, and F. Nebout, "Moddrop: Adaptive Multi-Modal Gesture Recognition," *IEEE Transactions on Pattern Analysis and Machine Intelligence*, Vol. 38, No. 8, 2015, pp. 1692–1706.

[38]   Zhao, M., T. Li, M. Abu Alsheikh, et al., "Through-Wall Human Pose Estimation Using Radio Signals," in *Proceedings of the IEEE Conference on Computer Vision and Pattern Recognition*, 2018, pp. 7356–7365.

[39]   Abavisani, M., H. R. V. Joze, and V. M. Patel, "Improving the Performance of Unimodal Dynamic Hand-Gesture Recognition with Multimodal Training," in *Proceedings of the IEEE Conference on Computer Vision and Pattern Recognition*, 2019, pp. 1165–1174.

[40]   Chen, V. C., *The Micro-Doppler Effect in Radar*, Norwood, MA: Artech House, 2019.

[41]   Aydogdu, C. Y., S. Hazra, A. Santra, and R. Weigel, "Multi-Modal Cross Learning for Improved People Counting Using Short-Range FMCW Radar," 2020 IEEE International Radar Conference (RADAR), Washington, DC, USA, 2020, pp. 250-255, doi: 10.1109/RADAR42522.2020.9114871.

[42]   Ester, M., H.-P. Kriegel, J. Sander, and X. Xu, "A Density-Based Algorithm for Discovering Clusters in Large Spatial Databases with Noise," in *KDD-96*, Vol. 96, No. 34, 1996, pp. 226–231.

[43]   Lin, T.-Y., P. Goyal, R. Girshick, K. He, and P. Dollar, "Focal Loss for Dense Object Detection,"in *Proceedings of the IEEE International Conference on Computer Vision*, 2017, pp. 2980–2988.

[44]   Santra, A., R. Santhanakumar, K. Jadia, and R. Srinivasan, "SINR Performance of Matched Illumination Signals with Dynamic Target Models," in *2016 IEEE International Conference on Acoustics, Speech and Signal Processing (ICASSP)*, 2016.

[45]   Ravanelli, M., and Y. Bengio. "Speaker Recognition from Raw Waveform with SincNet," in *2018 IEEE Spoken Language Technology Workshop (SLT)*, 2018, pp. 1021–1028.

9

Automotive In-Cabin Sensing

9.1 Introduction

In automotive solutions, radars play a pivotal role in enabling automation and autonomous driving. Short-range radars are preferred when building automotive solutions, such as collision warning where early warning is provided when the sides of the cars are too close to other objects, rear collision crash and blind spot monitoring for warning drivers of potential crash threats in blind spots and rear parking assist to ease the parking and navigation in areas such as parking lots [1–4] due to their low-cost and high resolution. The advent of deep learning has opened up the possibility of using these radars for a wider band of smart solutions. In this chapter, we will cover three distinctive solutions: (1) smart trunk opener (STO), (2) in-cabin vehicle occupancy sensing, and (3) feedback-aware distributed learning for vehicles.

In a vehicular system, STO enables the ease of opening or closing the trunk or door in a hands-free and key-free manner through a kick or hand gesture. Most STO solutions use capacitive sensors as proximity sensor to measure the closeness of an object but suffer from electromagnetic compatibility (EMC) disturbances and large false alarms are caused due to partial or entire covering up of the sensor with snow, dust, or water. Radars on the other hand can work in extreme weather conditions. However, for an entire working solution, a few more challenges need to be addressed, such as scenarios where a person walks too close to the sensor or vibrations due to luggage loading and people boarding, which may trigger the sensor as a false alarm. One of the solutions to these problems can be addressed by obtaining a Doppler spectrogram [5,6], which is a robust feature extraction and can be fed to a hidden Markov model (HMM) [7,8] machine learning classifier, which can act as a one vs. all classifier classifying a human kick motion over anything else, and thus can also suppress the intrinsic problem of unintentional triggering of STO due to presence or motion [9].

In-cabin vehicle occupancy sensing can help to develop a holistic autonomous driving solution that can be used for turning on seat heating, seat belt alarm detection, and also targeting air conditioning systems. In fact, in EURO NCAP child presence detection has been mentioned in its roadmap, instructing vehicle manufacturing companies to offer it as a feature addressing the issue of notifying the driver of children left inside the car [10]. One common approach is using a camera or a lidar for the purpose; however, it is not a well-accepted solution in terms of privacy and cost perspectives. The use of radars for this solution exhibits challenges such as multiple detections per object and yields very sparse data compared to the lidar or camera data. One way of addressing the following problem is to use radar data as point cloud taking into consideration the Doppler velocity and RCS values. Even though the radar data is sparse in nature, one possible way is to feed it to PointNets [11–13], a neural network approach that can perform 2D object detection and perform 3D instance segmentation [14] and 3D bounding box estimation to classify an adult/child and determine its occupancy.

Generally, developing end-to-end autonomous driving requires a lot of data, of which most comes from simulators due to limitations. However, a training model on data mostly generated from simulators may not work with high proficiency in the real world. This challenge can be addressed by acquiring data from all possible real-life environments or by performing collaborative learning. One can address the following problem by adapting a variant of federated learning (FL) [15,16]. FL has become one of the bases for distributed learning [17,18] and builds a relatively better collaborative solution that takes into account the feedback of the user about the currently deployed solution, retrains and personalizes the solution for the user while sending back only the weight difference for global learning, and also ensuring privacy of data. This method allows to build a more generalized solution over time with lower computational and latency overhead.

The chapter is laid out as follows: Section 9.2 presents the smart trunk opening solution, outlining its challenges, solutions, and results. In Section 9.3, we present the problem of vehicle occupancy sensing for child-left-behind applications and smart air bag solutions. In Section 9.4, we present the federated learning solution for vehicles, which is based on gesture-based distributed feedback to improve a general pedestrian classifier in vehicles. We conclude with some future work and direction in Section 9.5.

9.2 Smart Trunk Opening

9.2.1 Challenges

The current STO system used in vehicles is based on capacitive sensors, however there are several limitations and challenges associated with such sensors.

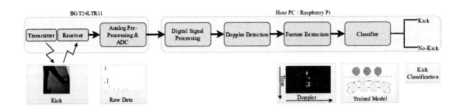

Figure 9.1 Block diagram of the radar STO system [9].

The capacitive sensors suffer from false alarms due to inability to distinguish between foot motion or a hovering animal, or EMC disturbances. Further, the performance of such sensors drops under wet or icy environments and when the sensor is covered with dust. In contrast, radar sensors offer an all-weather alternative solution while addressing several challenges faced from EMC disturbance. However, a signal processing based approach using short-range radars is still suboptimal and not product-ready since they rely on simple motion sensing and are triggered if people walk close to the door or trunk of the vehicle, and the system also gets triggered due to vibrations that are generated when passengers board the vehicle or luggage is boarded or de-boarded on the vehicle.

These challenges for short-range radars can be addressed through machine learning or deep learning approaches. One of the approaches uses extracting robust features such as Doppler spectrogram and fed into classical machine learning classifiers, and another approach uses the raw ADC time data into a deep neural network architecture for classification, which implicitly does the feature extraction and selection. We demonstrate a STO solution using *Infineon's BGT24LTR11* 24-GHz Doppler radar utilizing Doppler spectrogram as features followed by an HMM for differentiating valid or intentional kicks against an interfering Doppler signature from vibrations, people walking close to the trunk, or other spurious motion.

9.2.2 Solution

The STO is a one-class classification problem. One-class classification is encountered in many real-world applications such as novelty detection, anomaly detection, and authentication. Typically, such problems are solved in the framework of generic object classification. In our case, we address the problem of alien/other class by relying on robust feature extraction approaches, such as thresholded Doppler spectrogram, to improve the classification accuracy and choice of classifier, such as the HMM. HMM which relies on particular transition of emission and hidden states, offers a robust solution to our one-class classification problem. The proposed STO pipeline is depicted in Figure 9.1.

A short-time Fourier transform (STFT) is applied to the samples of ADC data per frame from the 24-GHz Doppler radar. The resultant kick Doppler-time spectrogram captures both the macro-Doppler component due to a person's leg movement during a kick toward the radar as well as the micro-Doppler components generated during the artifacts of the kick motion. The 2D-Doppler spectrogram is generated by STFT across 30 consecutive frames capturing evolution of Doppler spectrum over 6s with frame time of 200 ms. Examples of 2D-Doppler spectrogram for valid kicks and nonvalid kicks are shown in Figure 9.2 and Figure 9.3, respectively. As could be observed, kick motion can be defined by two states, one with an incoming Doppler (due to inward motion), and outgoing Doppler (due to outward motion) within specific Doppler limits. As presented in Figure 9.1, the radar echo is processed further to generate the Doppler spectrogram over 6s once the detected Doppler component crosses an adaptive threshold. Then, the extracted features are fed to the classification engine as described in following section for classifying the data to be kick or no-kick.

It is evident from the problem construct that the classifier would suffer from data asymmetry, since there would be numerous nonvalid kick Doppler spectrogram data compared to valid Doppler spectrogram data. Data imbalance can lead to biases in the classification, favoring the class with more data. In this approach, since we use the conventional machine learning algorithm, traditional approaches to handle class imbalance such as SMOTE [19] are applied on the training data set. The SMOTE algorithm equalizes class imbalance by oversampling minority classes (in this case kick data) with synthetic samples during training and thus preventing overfitting.

In Markov models, the states are directly visible to the observer and therefore the state transition probabilities are the only parameters, whereas in the hidden Markov model, the states are not directly visible, but the output observed feature vectors are visible, which are generated through emission probabilities depending on the hidden state. In the case of valid kick classification, the hidden states are modeled as inward motion, state 0, and outward motion, state 1. The sequence of feature vectors generated by HMM gives information about the sequence of states, and thus can be used to discern if a valid kick has been performed. HMM relies on the assumption that the ith hidden variable given the $(i - 1)$th hidden variable is independent of previous hidden variables, and the current observation variables depend only on the current hidden state. One of the classical approaches to estimate the parameters of HMM given observed feature vectors is the Baum-Welch algorithm [20], which uses the expectation-maximization (EM) algorithm. Figure 9.4 presents the HMM structure using two hidden states to model the kick vs. nonkick motion problem.

A HMM is a characterized as below:

$$\lambda = (A, B, \pi) \tag{9.1}$$

where λ is a model.

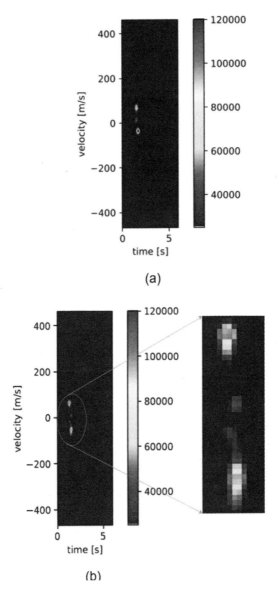

Figure 9.2 Doppler spectrogram of a valid kick (a), (b) from different people [9].

If S is a set of N states, and V is the observation set:

$$S = (s_1, s_2, ..., s_N), \qquad V = (v_1, v_2, ..., v_M) \tag{9.2}$$

Also, Q to be a fixed state sequence of length T, and corresponding observations O:

$$Q = (q_1, q_2, ..., q_T), \qquad O = (o_1, o_2, ..., o_T) \tag{9.3}$$

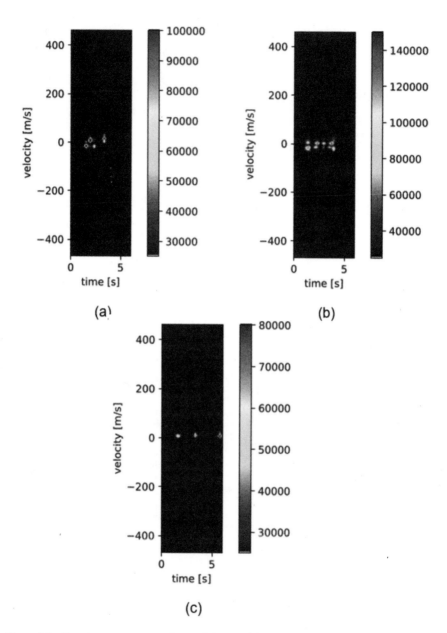

Figure 9.3 Doppler spectrogram of nonvalid kicks: (a) hand movements, (b) vibration, and (c) walking [9].

A is a $N \times N$ transition matrix, which contains the probability of state j following state i. The state transition probabilities are independent of time t:

$$A = [a_{ij}], \quad a_{ij} = P(q_t = s_j | q_{t-1} = s_i) \tag{9.4}$$

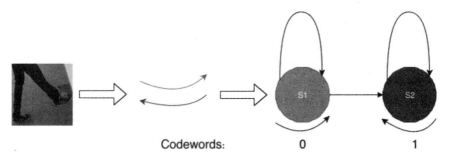

Figure 9.4   Hidden Markov model with two hidden states representing the inward and outward motions of the kick [9].

B is a *NXN* observation matrix, which contains the probability of observation k being produced from the state j, independent of time t:

$$B = [b_i(k)], \qquad b_i(k) = P(x_t = v_k | q_t = s_i) \tag{9.5}$$

π is the initial probability array:

$$\pi = [\pi_i], \qquad \pi_i = P(q_1 = s_i) \tag{9.6}$$

The model presumes:

1. *Markov property:* The current state depends only on the previous state (the memory of the model); that is, $P(q_t | \{q\}_1^{t-1}) = P(q_t | q_{t-1})$.
2. *Assumption of independence:* The output observation at time t is dependent only on the current state, it is independent of previous observations and states; that is, $P(o_t | (\{o\}_1^{t-1}, \{q\}_1^t)) = P(o_t | q_t)$.

During the training, the parameters (A, B, π) are adapted in order to obtain the model that best narrates the spatiotemporal dynamics of the desired gesture by optimizing the maximum likelihood measure $\log(P(\text{observation}|\text{model}))$ over a set of training examples. This optimization process requires expensive computation methods. The HMM states are hidden.

The three fundamental aspects of HMM are:

1. *Evaluation:* Calculating the likelihood of the data when the model parameters and observed data are given;
2. *Decoding:* Estimating the optimal sequence of hidden states when the model parameters and observed data are given;
3. *Training:* Estimating the model parameters when the observed data is given.

The three aspects can be solved using forward-backward, Viterbi, and an iterative EM algorithm, known as the Baum-Welch (BW) algorithm, respectively.

9.2.3　Results and Discussion

The kick data set is recorded using *Infineon's BGT24LTR11* Doppler radar chipset by placing the radar sensor 45–55 cm above the ground on a tripod (to emulate the placement on a vehicle's trunk). Kick Doppler spectrograms from 15 individuals were recorded by asking them to make kicks toward the trunk sensor from various directions (i.e., left, center, and right). The recording script starts recording as a motion is detected through the adaptive thresholding mechanism. The nonvalid kick data is recorded by walking close to the radar sensor in both directions, and the vibration data is induced by manually vibrating the sensor to emulate the effect of a passenger sitting in the seats. Random objects were also moved close to the sensor, including hand movements to generate the nonkick data set. For each recording, the Doppler spectrogram is built by collecting 30 consecutive frames and 128 ADC samples. The radar transmit power is limited to 6 dB to ensure the received radar signal is not saturated. Table 9.1 presents the system parameters used for recording the Doppler spectrogram. The training data set comprises a total of 480 kick data and 800 nonkick data, while the test data set contains 60 kick and nonkick data.

The HMM model is trained using the Baum-Welch algorithm on the training data set. The performance of the trained classifier on the test data during inference resulted in 99.17% classification accuracy, as presented in Figure 9.5. Table 9.2 summarizes the accuracy result and the model size.

Furthermore, similar algorithm can be applied for a smart rooftop opener, which involves moving the fingers close to the sensor and away within a margin of time duration. The concept of a HMM with two states of positive Doppler followed by negative Doppler could be directly deployed for similar automotive in-cabin solution as well. Furthermore, the concept of valid kick sensing can be deployed in industrial and consumer waste bins, where users can open/close the lid of the waste bin through kick without having to physically touch them.

9.3　Vehicle Occupancy Sensing

9.3.1　Challenges

The need for sensors with smart sensing capabilities to detect and classify children in vehicles has become mandatory after the EURO NCAP regulations. Furthermore, the detection and identification of passenger seat occupancy enables controlled firing of the airbags. Smart airbags rely on smart sensing to sense information about occupant position, size, and posture to discriminate between adults, children, front- or rear-faced child seats, objects put on the seat, or simply empty seats [21].

Using short-range radars, one of the potential solutions relies on vital sensing in combination with vital Doppler sensing to identify if a human or nonhuman is occupying the passenger seat. However, such a solution is not enough for the above

Table 9.1
System parameters [9]

Parameters	Symbol	Value
Operational frequency	f	24 GHz
Number of samples/snapshot	N_s	128
Frame time	T_f	200 ms
Number of frames	N_f	30
Doppler resolution	V_r	7.32 m/s
Maximum velocity	V_{max}	461.41 m/s

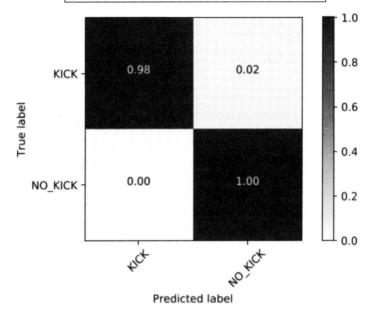

Figure 9.5 Normalized confusion matrix [9].

Table 9.2
Summary of the classification results [9]

Method	Accuracy (in %)	Model size
HMM	99.17	248KB

two applications, smart airbags and child-left-behind, for distinguishing a child vs. adult based on their size and dimension. Furthermore, if the vehicle's engine is on, the vibration arising from the engine is enough to sway over the vital-Doppler motion, thus a solution based on vital sensing isn't a viable solution for such a problem. One solution to the problem involves generating two-dimensional spatial imaging across range bins through vital-Doppler data frames after zero-Doppler subtraction, followed by either 3D detection and 3D clustering through

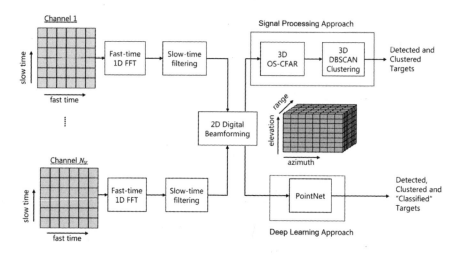

Figure 9.6   Processing pipeline to generate radar 3D point cloud for occupancy sensing followed by either a signal processing or deep learning approach for analysis.

a signal processing approach or classification of point clouds through PointNet architecture, as presented in Figure 9.6. The $N_{tx} \times N_{rx}$ filtered range spectrum are stacked into a matrix along elevation ϕ and azimuth θ as per the position of the virtual antennas. The effective array factor matrix $w(\theta, \phi)$ is calculated as the Kronecker product of the steering vector of the Tx array $a_{Tx}(\theta, \phi)$ and the steering vector of the Rx array $a_{Rx}(\theta, \phi)$ and used to calculate the target angles $(\theta_t, \phi_t)_{t=1}^{T}$ for all range bins, where T are all the target scatterers. The angle profiles for each range bin along θ and ϕ are generated through either a Capon beamformer or maximal likelihood estimation. Figure 9.7(a) and (d) illustrates 3D point cloud along range bins, azimuth, and elevation angle for an adult and child, respectively. PointNets work on 3D point clouds utilizing the size and volume of point clouds to distinguish between adult and child, and front- or rear-faced child seats, and objects put on the seat.

9.3.2 Solution

The received baseband signal from the kth target scatterer can be expressed as

$$\bar{s}_{\text{RX}}(t) = \rho_k e^{\frac{-j2\pi 2u_k \cdot r_k}{\lambda}} a_n^{Rx}(\theta_k, \phi_k) a_m^{Tx}(\theta_k, \phi_k)^T \bar{s}_{\text{TX}}(t) \qquad (9.7)$$

where ρ_k represents the composite amplitude contribution due to propagation path loss, antenna gains, and receiver gains. $\bar{s}_{\text{TX}}(t)$ and $\bar{s}_{\text{RX}}(t)$ are the transmitted signals from N_{TX} transmit antennas and the received signal at N_{RX} receive antennas, respectively. $a_m^{Tx}(\theta_k, \phi_k)$ and $a_n^{Rx}(\theta_k, \phi_k)$ represent the transmit and receive steering vector along azimuth θ and elevation angle ϕ. u_k and r_k are the unit

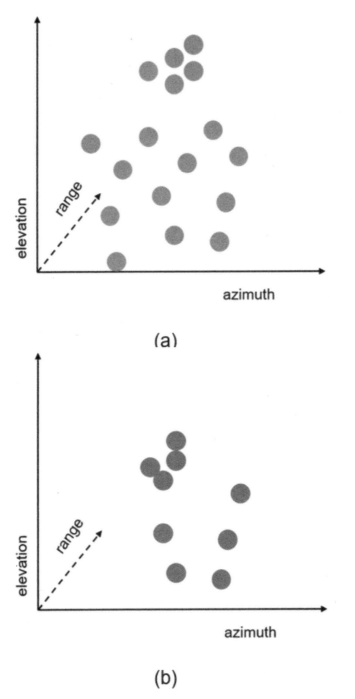

Figure 9.7 Illustration of radar point cloud across range, elevation, and azimuth from sensed (a) adult, and (b) child.

vector toward the target direction and the range directional vector, respectively. After having estimated the range R_k of the kth target through spectral analysis along fast time, the point cloud spread across the angular coordinates of the target, azimuth angle θ and elevation angle ϕ, are generated using 2D Capon algorithms, resulting in point clouds such as the ones in Figure 9.7.

Given radar point clouds from the 2D Capon algorithm, the objective of the 3D deep neural network is to classify and segmentize objects in 3D space (Figure 9.7). The radar point cloud is represented as a set of three-dimensional points $p_k = (r_k, \theta_k, \phi_k)$, $k = 1, \ldots, K$, where K are the number of detected target points. For the classification task, the objective is to distinguish between child, adult, luggage, or empty.

To ensure the invariance under geometric transformation (i.e., point cloud rotation shouldn't alter the classification or segmentation results), a transformation T-Net is applied to transform the input feature vectors into

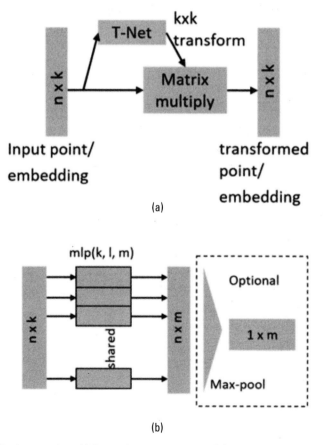

Figure 9.8   Basic structure of (a) transformation net, and (b) vanilla PointNet comprising the PointNet for classification and segmentation tasks.

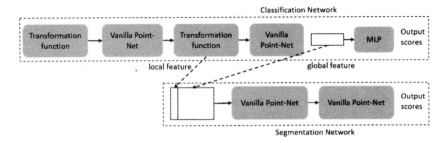

Figure 9.9 High-level example of a PointNet based on the original PointNet architecture.

transformed feature vectors. This operation is illustrated in Figure 9.8(a) where n points with k dimension is applied through T-Net learning transform parameters, $k \times k$, which can be applied through matrix multiplication on the input feature vectors, resulting in output/transformed feature vectors, $n \times k$.

Further, by the universal set function approximation theorem, a Hausdorff continuous symmetric function $f : 2^x \in \mathcal{R}$ can be arbitrarily approximated by PointNet;

$$|f(s) - \gamma(\max_{x_i \in S} h(x_i))| < \epsilon \qquad (9.8)$$

where ϵ is an arbitrary small number. The above equation is what the vanilla PointNet tries to replicate, by implementing the $\gamma(.)$ and $h(.)$ functions are implemented using MLP and max(.) operation is executed by the max-pooling. The vanilla PointNet implementing these operations, as proposed in the original PointNet paper, are illustrated in Figure 9.8(b). Figure 9.9 presents an example architecture of PointNet, which uses a combination of a series of transformation function (through T-Net) and vanilla PointNet inspired from the original PointNet architecture. The transformation function at the input makes the input data points invariant to geometric transformations, while at intermediate layers makes the input embedding vectors invariant to geometric transformations. The network architecture has both the classification network and the segmentation network. The segmentation part of the network marks each pixel and which class it belongs to, and it takes input from local and global features followed by a sequence of vanilla PointNets.

9.4 Federated Learning

9.4.1 Challenges

Deep learning (DL) is one of the most widely used technologies in developing intelligent automotive solutions and autonomous driving. In general, a multistep process is followed for developing and deploying deep learning solutions, specifically for automotive use. This involves the step of pretraining DL models on simulator based data due to limited availability of automotive data. These pre-trained DL models are then deployed in real-life scenarios and the models are fine-tuned for the local device. However, the entire multistep process is extremely time consuming, which directly affects production time. The bigger problem of such a process is that the fine-tuned models remain locally and there is no collaborative learning, which is a must-have due to the high mobility of vehicles, which makes it subject to huge variability in environment and user interaction.

An alternative solution to address the above problems is having a centralized storage, where the initial model is deployed across multiple devices and new data is collected from all the devices and stored in the central storage for retraining a more generalized model that is updated across these devices. However, this solution poses two major problems of very high latency due to the data transfer and privacy specifically when used in automotive solutions. Recently a new training method FL, has become the basis for privacy-aware distributed learning. To have a simpler understanding of the training method, let us take up the use case of a music recommender system using FL. Initially a copy of the model (global model) is deployed on every device and the model is continuously trained locally to personalize to the musical preferences of specific users. The changes in weights of the model are sent back to the central server where the new learnings are aggregated and the global model is updated and a copy is sent back to the devices. This phenomenon is iterative, which results in an improved model over time. Inspired by the idea of federated learning, we propose a feedback-aware distributed learning methodology. To adhere to the scope of the book, we fit our learning methodology for vision-based object classification used in automotive

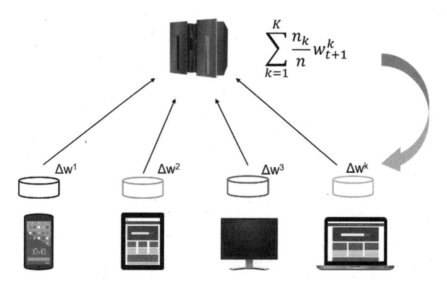

$$\sum_{k=1}^{K} \frac{n_k}{n} w_{t+1}^k$$

Δw^1 Δw^2 Δw^3 Δw^k

Figure 9.10 General federated learning architecture.

solutions for pedestrian/nonpedestrian classification and use our radar based gesture sensing solution for enabling gesture-based feedback. Figure 9.10 presents a general federated learning architecture.

9.4.2 Solution

Initially a copy of the pretrained global vision model performing the task of the pedestrian/nonpedestrian object classification with a certain degree of accuracy and precision is deployed across multiple devices. Since there is a huge variability among the walk patterns, posture, and so on, among pedestrians across different environments, we would like the local model to adapt and learn from such local artifacts as well as this learning to be able to enrich the global model collaboratively while restricting transfer of data from the local device. We propose to use radar-based in-cabin gesture sensing to provide feedback about the performance of the object classifier due to the ease of use, especially during driving. The user may opt to provide a gesture feedback after every classification done by the object classifier. For simplification, let us define two possible gestures, positive and negative. Figure 9.11 illustrates the general positive and negative feedback.

Now based on the feedback given for every classification, we label the newly collected data locally as the predicted class if it was a positive feedback or the opposite of the predicted label if it was a negative feedback. Once we have a fixed amount of newly labeled data, the next step is to fine-tune the global model locally. However, retraining the entire object classification model can be computationally heavy and may increase latency due to transfer of the entire model weight to the server for its participation in distributed learning. In practice, the weight of initial layers of a model are frozen and only the last few layers take part in retraining while fine-tuning the model. This makes the process relatively computationally inexpensive and low latency for transfer.

Since the local model will eventually take part in improving the global model, we must ensure that there is no mislabeled data due to malicious intentions or unintentional gestures. To ensure it we make sure that the prediction confidence metric is below a defined threshold, given a negative feedback, and vice versa for a positive feedback. After a given period of time, local devices with an updated local model participate in the collaborative learning, where neither the newly labeled

Figure 9.11 Gesture-based feedback for model performance.

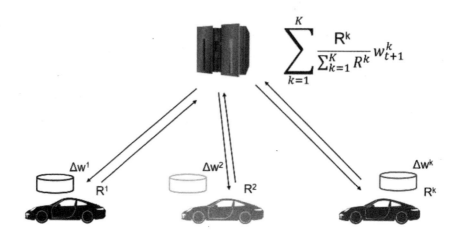

Figure 9.12   A proposed federated learning-based feedback mechanism for an automotive classification model.

local data nor the updated local model is transferred to the global server but the weight differences of the last few layers that took part in fine-tuning and that of the copy of the last global model are sent to the global server along with the total number of negative feedbacks. The number of negative feedbacks allows us to perform weighted averaging to boost their significance in new learning. The output of the averaged weights are added to the layers that took part in fine-tuning and the updated global model is dispatched across all the local devices. Figure 9.12 illustrates the proposed federated learning-based infrastructure for updating an automotive classification model.

9.5   Future Work and Direction

The smart trunk opening solution presented in the chapter needs improvement in terms of power savings to enable adaption in low-cost, low-power automotive solutions. The PointNet solution presented to segment and classify adult, child, luggage, or empty space can be further improved by using a neural network to also generate the 3D point clouds, instead of generating them through the 2D Capon algorithm presented here.

9.6   Problems

1. What are some applications of deep learning based radar solutions in vehicles?

2. What are some possible disadvantages of feeding raw ADC data to classical machine learning classifiers?
3. Describe a deep learning approach to build a STO. Discuss the input data, architecture, training approach, and loss function.
4. Why are vital sensing based approaches not a good choice for vehicle occupancy sensing?
5. What is the use of T-Net in PointNet architecture?
6. What are the specific advantages of federated learning?
7. What is the importance of counting negative feedback in the described federated learning approach?
8. What problem can arise if all the clients in the described federated learning approach mostly see nonpedestrian objects? How can it be tackled?
9. Explain the other parameters, apart from the point clouds, that can be used for adult-child classification in a radar PointNet.

References

[1] Hasch, J., E. Topak, R. Schnabel, T. Zwick, R. Weigel, and C. Waldschmidt, "Millimeter-Wave Technology for Automotive Radar Sensors in the 77 GHz Frequency Band," *IEEE Transactions on Microwave Theory and Techniques*, Vol. 60, No. 3, 2012, pp. 845–860.

[2] Patole, S. M.. M. Torlak, D. Wang, and M. Ali, "Automotive Radars: A Review of Signal Processing Techniques," *IEEE Signal Processing Magazine*, Vol. 34, No. 2, 2017, pp. 22–35.

[3] Jones, W. D., "Keeping Cars from Crashing," *IEEE Spectrum*, Vol. 38, No. 9, 2001, pp. 40–45.

[4] Meinel, H. H., "Evolving Automotive Radar: From the Very Beginnings into the Future," in *The 8th European Conference on Antennas and Propagation (EuCAP 2014)*, 2014, pp. 3107–3114.

[5] Chen, V. C., F. Li, S.-S. Ho, and H. Wechsler, "Analysis of Micro-Doppler Signatures," *IEE Proceedings-Radar, Sonar and Navigation*, Vol. 150, No. 4, 2003, pp. 271–276.

[6] Chen, V. C., F. Li, S.-S. Ho, and H. Wechsler, "Micro-Doppler Effect in Radar: Phenomenon, Model, and Simulation Study," *IEEE Transactions on Aerospace and Electronic Systems*, Vol. 42, No. 1, 2006, pp. 221.

[7] Rabiner, L. R. "A Tutorial on Hidden Markov Models and Selected Applications in Speech Recognition," *Proceedings of the IEEE*, Vol. 77, No. 2, 1989, pp. 257–286.

[8] Elmezain, M., A. Al-Hamadi, J. Appenrodt, and B. Michaelis, "A Hidden Markov Model-Based Continuous Gesture Recognition System for Hand Motion Trajectory," in *2008 19th International Conference on Pattern Recognition*, 2008, p. 14.

[9] Shankar, Y., and A. Santra, "Valid Kick Recognition in Smart Trunks based on Hidden Markov Model Using Doppler Radar," *2019 International Radar Conference (RADAR)*, IEEE, Toulon, France, September 23–26, 2019.

[10] Santra, A., R. V. Ulaganathan, and T. Finke, "Short-Range Millimetric Wave Radar System for Occupancy Sensing Application," *IEEE Sensors Letters*, Vol. 2, No. 3, 2018, p. 14.

[11] Qi, C. R., H. Su, K. Mo, and L. J. Guibas, "PointNet: Deep Learning on Point Sets for 3DClassification and Segmentation," *Proc. Computer Vision and Pattern Recognition (CVPR)*, IEEE, 2017.

[12] Qi, C. R., L. Yi, H. Su, and L. J. Guibas, "PointNet++: Deep Hierarchical Feature Learning on Point Sets in a Metric Space," *Advances in Neural Information Processing Systems*, Vol. 30, 2017, pp. 5099–5108.

[13] Qi, C. R., W. Liu, C. Wu, H. Su, and L. J. Guibas, "Frustum Pointnets for 3DObject Detection from RGB-D Data," *Proc. Computer Vision and Pattern Recognition (CVPR)*, IEEE, 2018.

[14] Lombacher, J., K. Laudt, M. Hahn, J. Dickmann, and C. Wohler, "Semantic Radar Grids," in *IEEE Intelligent Vehicles Symposium (IV)*, June 2017, pp. 1170–1175.

[15] Yang, Q., Y. Liu, T. Chen, and Y. Tong, "Federated Machine Learning: Concept and Applications," *ACM Transactions on Intelligent Systems and Technology (TIST)*, Vol. 10, No. 2, 2019, p. 12.

[16] Pan, S. J., and Q. Yang, "A Survey on Transfer Learning," *IEEE Transactions on Knowledge and Data Engineering*, Vol. 22, No. 10, 2009, pp. 1345–1359.

[17] Liu, B., L. Wang, M. Liu, and C. Xu, "Lifelong Federated Reinforcement Learning: A Learning Architecture for Navigation in Cloud Robotic Systems," arXiv preprint arXiv:1901.06455, 2019.

[18] Nadiger, C., A. Kumar, and S. Abdelhak, "Federated Reinforcement Learning for Fast Personalization," in *2019 IEEE Second International Conference on Artificial Intelligence and KnowledgeEngineering (AIKE)*, IEEE, 2019, pp. 123–127.

[19] Chawla, N. V., K. W. Bowyer, L. O. Hall, and W. P. Kegelmeyer, "SMOTE: Synthetic Minority Oversampling Technique," *Journal of Artificial Intelligence Research*, Vol. 16, 2002, pp. 321–357.

[20] Bishop, C. M., "Pattern Recognition and Machine Learning," *Information Science and Statistics*, 2006.

[21] Fritzsche, M., C.Prestele, G. Becker, M. Castillo-Franco, and B.Mirbach, "Vehicle Occupancy Monitoring with Optical Range-Sensors," *2004 IEEE Intelligent Vehicles Symposium*, Parma, Italy, June 14–17, 2004.

About the Authors

Avik Santra received his B.E. degree from the West Bengal University of Technology in 2008, and M.E. degree (Hons.) in signal processing from the Indian Institute of Science, Bengaluru, in 2010. Early in his career, he was a system engineer developing signal processing algorithms for LTE/4G modem chipsets at Broadcom Communications. Subsequently, he worked at Airbus as a research engineer developing algorithms for artificial-intelligence-based radars. He is currently senior staff expert algorithm engineer developing signal processing and machine learning algorithms for industrial, consumer, and automotive short-range radars and depth sensors at Infineon Technologies AG, Neubiberg, Germany. He is a Senior Member of IEEE. He has filed over 40 patents and has published over 30 research papers related to various topics of radar waveform design, radar signal processing, and radar machine/deep learning. He is a reviewer for various IEEE and Elsevier journals, and was the recipient of several outstanding reviewer awards.

Souvik Hazra received his B.Tech. degree from KIIT University in 2017 and M.S. degree in data science and engineering from EURECOM & IMT, France, in 2019. He has worked in the field of deep learning with organizations like University of Cambridge and Airbus, and is currently working as an AI engineering consultant for Infineon Technologies AG, Neubiberg, Germany. He currently works in developing machine learning and deep learning solutions with industrial, consumer, and automotive short-range radars and microphones. He has so far filed 5 patents and has published over 10 research articles on machine learning and deep learning topics. He is also working toward his PhD in multimodal deep learning from Friedrich-Alexander-University Erlangen, Germany, in his spare time.

Index

Recent Titles in the Artech House
Radar Series

Space-Time Adaptive Processing for Radar, Second Edition,
  Joseph R. Guerci

Special Design Topics in Digital Wideband Receivers, James Tsui

Systems Engineering of Phased Arrays, Rick Sturdivant, Clifton Quan,
  and Enson Chang

Theory and Practice of Radar Target Identification, August W.
  Rihaczek and Stephen J. Hershkowitz

Time-Frequency Signal Analysis with Applications, Ljubiša Stanković,
  Miloš Daković, and Thayananthan Thayaparan

Time-Frequency Transforms for Radar Imaging and Signal Analysis,
  Victor C. Chen and Hao Ling

Transmit Receive Modules for Radar and Communication Systems,
  Rick Sturdivant and Mike Harris

For further information on these and other Artech House titles, including previously considered out-of-print books now available through our In-Print-Forever® (IPF®)
program, contact:

Artech House	Artech House
685 Canton Street	16 Sussex Street
Norwood, MA 02062	London SW1V HRW UK
Phone: 781-769-9750	Phone: +44 (0)20 7596-8750
Fax: 781-769-6334	Fax: +44 (0)20 7630-0166
e-mail: artech@artechhouse.com	e-mail: artech-uk@artechhouse.com

Find us on the World Wide Web at: www.artechhouse.com